Aviation Safety and Security

Aviation Safety and Security

Utilising Technology to Prevent Aircraft Fatality

Stephen J Wright

CRC Press
Taylor & Francis Group
Boca Raton London New York

CRC Press is an imprint of the
Taylor & Francis Group, an **informa** business

First edition published 2021
by CRC Press
6000 Broken Sound Parkway NW, Suite 300, Boca Raton, FL 33487-2742

and by CRC Press
2 Park Square, Milton Park, Abingdon, Oxon, OX14 4RN

CRC Press is an imprint of Taylor & Francis Group, LLC

ISBN: 978-0-367-27519-8 (hbk)
ISBN: 978-1-032-01344-2 (pbk)
ISBN: 978-0-429-29645-1 (ebk)

Typeset in Palatino
by SPi Global, India

Contents

Preface...ix
Acknowledgments..xi
Author.. xiii
List of Abbreviations..xv

1. Introduction 1

2. History of Security Events 11
   2.1   Introduction 11
   2.2   Threat and the Use of Violence in Aviation..............................11
   2.3   Murder at the Controls of an Aircraft 13
   2.4   The Second World War and the Immediate Period Thereafter 14
   2.5   The Late 1950s and 1960s..16
   2.6   The 1970s Onwards...17
   2.7   Conclusions..21

3. Flight Data Recorders and Cockpit Voice Recorders...........................23
   3.1   Introduction 23
   3.2   Flight Instrument Recording 25
   3.3   Certification and Flight Instrument Recording 27
   3.4   Decoding Flight Instrument Data from Data Recorders 30
   3.5   Conclusions..31

4. Flight Controls and Environmental Control Systems 33
   4.1   Relevance of Flight Controls Air-Conditioning Systems and
         Commercial Aviation..33
   4.2   Comparing the Underlying Philosophy of Flight Controls
         for Airbus and Boeing 34
   4.3   Air-Conditioning Systems and Commercial Aviation.................38
   4.4   Conclusions..48

5. Use of Live Aircraft Data in Aircraft Maintenance Management.........51
   5.1   Introduction 51
   5.2   Aircraft Maintenance Management and Its Commercial
         Importance 51
         5.2.1   Maintenance Planning by the Original Equipment
                 Manufacturer (OEM)..52
         5.2.2   Unscheduled Maintenance.......................................53

	5.2.3	Minimum Standards of Equipment of Systems – Master Minimum Equipment List	53
	5.2.4	Component Reliability and Maintenance Strategy	54
	5.2.5	Bathtub Curve for Reliability and Mathematical Predictions	55
5.3		Technical Components Combined with Data Logging	57
5.4		Live Streamed Data and Radio Communication Technologies	61
	5.4.1	ACARS	61
	5.4.2	British Airways Engineering Maintenance Management	64
5.5		Data Mining of Very Large Data Bases and Commercial Solutions to Predictive Maintenance	64
	5.5.1	Rolls Royce Commercial Engines	64
	5.5.2	Airbus, Palantir and Skywise	67
5.6		Conclusions	69

6. Human Factors and Safety Management Systems 71
6.1	Relevance of Human Performance and Safety Management Systems	71
6.2	British European Airways Accident – A Turning Point	72
6.3	SHELL Model	76
6.4	*The Impossible Accident* – Tenerife, 1977	78
6.5	Error Chain Model	81
6.6	Flight Crew Training to Prevent Events	83
6.7	Professor James Reasons' *Swiss Cheese Model*	85
6.8	Safety Management Systems	89
6.9	Conclusions	91

7. Aircraft Security 93
7.1	Introduction	93
7.2	Flight Decks with Curtains and Doors	94
7.3	British Airways 2069, 29th December 2000	95
7.4	Year 2001 Events and Post 9-11 Modifications to the Flight Deck Door	98
7.5	Conclusions	103

8. Unusual Losses of Aircraft 105
8.1	Introduction	105
8.2	Korean Air Lines KL 007 Shoot Down, 1 September 1983	105
8.3	Pacific South West Airlines Flight 1771, 7 December 1987	109
8.4	Silk Air 185, 19 December 1997	111
8.5	Egypt Air 990, 31 October 1999	114
8.6	Malaysian Airlines MH370, 8 March 2013	117

8.7   LAM Mozambique Flight 470, 29 November 2013 135
8.8   Germanwings 9525, 24 March 2015 136
8.9   Horizon Air (Theft/Suicide), 10 August 2018 142
8.10   Other Current Sources of Weaknesses in Commercial
       Aircraft 145
       8.10.1 Aircraft Toilet Smoke Detectors 146
       8.10.2 Viral Contamination and Cross Infection within
               the Aircraft 148
8.11   Conclusions 150

9. **Minimizing Loss: Modifying Current Aircraft and Processes** 153
9.1   Introduction 153
9.2   History of Remotely Controlled Aircraft 154
9.3   Federal Aviation Administrations' Full-Scale Controlled
       Impact Demonstration 158
9.4   Remote Controlled Aircraft (Drones) during and after the
       Gulf War – Operation Desert Storm 160
9.5   Boeing Patent on Remote Control Takeover of Aircraft 164
9.6   Current Capabilities and Their Limitations 167
9.7   Changes and Technologies Required for a Safe Autonomous
       System 169
       9.7.1 ECS Life Support 169
       9.7.2 Circuit Breakers 170
       9.7.3 Transponders 170
       9.7.4 One Time Use Codes 171
       9.7.5 Satellite Communication Uplinked Continuous Data 171
9.8   The Justification and Driver to Introduce Ground Monitored
       Technologies 172
       9.8.1 Financial Drivers 173
       9.8.2 Live Streamed Data Reducing the Fuel Burn 174
       9.8.3 Live Streamed Data Reducing Deviation from Flight
              Plans and Further Reducing Fuel Burn 174
       9.8.4 Data Security and Encryption for Uninterrupted
              Landing Systems 175
       9.8.5 Uninterrupted Landings and Risks Posed from the
              Aircraft's Occupants 176
9.9   Conclusions 178

Index 183

Preface

The author has written this book because the general subject of safety and security in the airline industry is usually a discipline that many airlines claim to be in their top commitments, but little is explained as to how the industry meets the claimed achievement. From an academic perspective, when looking through the published materials, while there are numerous publications that detail the more common list of security events, very few explain what developed within the industry as a result of various incidents, including the changes in legal requirements. For example, post-9/11, flight deck doors were identified by the Original Equipment Manufacturers (i.e. Boeing/Airbus) as the strategic point of weakness, and were thus modified to prevent forced entry from the cabin side of the door. However, there have been numerous in-flight security events where the occupants of the flight deck have taken advantage of such changes, preventing access to the flight deck prior to a catastrophic crash.

Likewise, in-flight data recording of flight parameters and noises/voices inside the flight deck were designed to capture at least the last 30 minutes of the flight prior to a crash. While the explanation of development of these devices is common knowledge, the modern-day use of this captured data from the 'black boxes' is not well explained, and the recording of multiple systems gives engineers on the ground technical insight into how the aircraft is performing, to allow for better planning of maintenance activities. This application of technology can be further extended if the aircraft were to live stream all of the component data via a satellite communication service, allowing for real-time maintenance predictions. Such modifications would also enable Search And Rescue operations to become very precise, to immediately identify where the aircraft is located (from the last transmitted GPS position – longitude/ latitude) and to prevent extended searches for aircraft whose final resting place is unknown at the time of writing (e.g. MH370).

A further consideration is given to the importance of recording information in-flight, if all data can be monitored from the ground. Post catastrophic events (e.g. an aircraft crash), effectively, the 'black boxes' would no longer be the sole source of much of the information from within the aircraft, because all the data would have been transferred via satellite data to a ground-based server.

Lastly, the author hoped that further modifications and changes to enhance safety and security would be considered, such as changing protocols that pilots follow in-flight. For example, the Environmental Control System provides fresh air to all the passengers in the aircraft in-flight. The hypothetical question is raised, should a pilot (from a secure locked flight deck) be able to turn off all the cabin supplies of fresh air and furthermore depressurise the aircraft: the result of such actions would be unthinkable, but possible in the current technical climate. Should all critical life support systems be controlled from within the confines of a reinforced flight deck, or should certain systems only be deactivated when the aircraft lands with 'weight on wheels' detected.

Acknowledgments

I would like to thank my family for their support while I have been hidden away writing this book. My wife, Eun Kyeong, has been inspirational, patient and loving as our journey has taken us across the globe. To my daughter Cassie, I appreciate her shared interest and passion for aircraft, from building plastic models to long haul flights on Jumbo jets!

I would also like to thank my parents for their unconditional support and encouragement to make my own career choices. As a young child, all those Airfix aircraft and balsa wood aircraft models that I flew in the garden had a lasting effect on me.

Many thanks are given to my parents-in-law, who have always seen the positive in our family's aviation-themed journey.

Special thanks are given to Alexander Leece at the University of Leeds, for his diligent efforts to proofread the final manuscript.

Lastly, a special thanks is given to Taylor & Francis Publishers and Kyra, for their support and help in bringing this book to print.

Author

Stephen J Wright is a professorial academic member of staff in the Faculty of Engineering and Natural Sciences at Tampere University, Finland, specialising in aviation, aeronautical engineering and aircraft systems. In addition to his university activities, Prof. Wright continues (at the time of this writing) to hold the esteemed post of President for the Finnish Society of Aeronautical Engineers, is a past Member of the Royal Aeronautical Society, UK and is a lifelong Fellow of the Higher Educational Academy, UK.

Prof. Wright earned a PhD at the University of Leeds, UK, in the fouling and failure of commercial aircraft air-conditioning systems. Other academic qualifications include a postgraduate certification of teaching and learning in higher education, awarded by Kingston University, London. He earned a BSc in Chemistry at the University of Sussex, with additional ERASMUS studies at Uppsala Universitiet, Sweden.

He engages formally with numerous elements of the European Commission as a recognised expert in aviation/aeronautical engineering. He is fully committed to the objectives and aspirations to improve the industry that will allow for better and more effective air transportation (Flightpath 2050/Master Plan).

Prof. Wright has attended and passed the examinations for numerous manufacturer 'line and base' maintenance engineering-type rating courses (B1.1 discipline) associated with his previous commercial aviation employment, in addition to holding a European Aviation Safety Agency flight crew license for single-engine piston aircraft.

List of Abbreviations

AAIB	Air Accident Investigation Branch (UK)
ACARS	Aircraft Communication Addressing and Reporting System
ADC	Air Data Computer
ADD	Acceptable Deferred Defect
ADS-B	Automatic Dependent Surveillance – Broadcast
AES	Advanced Encryption Standards
AMM	Aircraft Maintenance Manual
AOG	Aircraft On Ground (sic *grounded*)
ATC	Air Traffic Control
ATTOL	Autonomous Taxi, Take-off and Landing system (Airbus)
BALPA	British Airline Pilot Association
BDD	Base Deferred Defect
BEA	Bureau d'Enquêtes et d'Analyses (Regulator, France)
BEA	British European Airways (precursor to British Airways Plc)
BOAC	British Overseas Airways Corporation (precursor to British Airways Plc)
CA	Cambridge Analytica
CAA	Civil Aviation Authority
Capt	Captain
Cat	Category (landing system number)
C/B	Circuit Breaker
CCC	Cockpit Coordinated Concept
CCD	Charged Coupled Device
CDA	Continual Descent Approaches
COVID 19	Severe Acute Respiratory Syndrome Coronavirus 2 (SARS-CoV-2)
CRM	Crew Resource Management
CVR	Cockpit Voice Recorder (analogue)
DCVR	Digital Cockpit Voice Recorder
DFDR	Digital Flight Data Recorder
DH	Decision Height
dP	Differential Pressure
E2EE	End 2 End Encryption
EADS	European Aeronautic Defence and Space (Company)
EASA	European Union Aviation Safety Agency
ECS	Environmental Control System
EGT	Exhaust Gas Temperature
ETS	Emissions Trading Systems (EU)
EWS	Enterprise-Wide System (BA IT system)

FAA	Federal Aviation Administration
FADEC	Full Authority Digital Engine Control
FAR	Federal Aviation Requirements (USA regulations)
FBI	Federal Bureau of Investigation
FDR	Flight Data Recorder (analogue)
FCOM	Flight Crew Operating Manual
FMC	Flight Management Computer
F/O	First Officer
GPS	Global Positioning System (*satellites*)
GPWS	Ground Proximity Warning Systems
GSM	Global System for Mobile (communications)
HF	Human Factors
INS	Inertia Navigation Systems
ICAO	International Civil Aviation Organisation
ILS	Instrument Landing System
Kts	Knots
LBA	Luftfahrt Bundesamt (NAA Germany)
LRU	Line Replaceable Unit
MEL	Minimum Equipment List
MMEL	Master Minimum Equipment List (OEM)
MTBO	Mean Time Between Overhaul
MTFF	Mean Time To Failure
MTOM	Maximum Take-Off Mass
NAA	National Aviation Authority
NTSB	National Transportation Safety Board (USA)
OEM	Original Equipment Manufacturer
PA	Pressure Altitude
PF	Pilot Flying (i.e. P1)
PFHE	Plate Fin Heat Exchanger
PNF	Pilot Not Flying (i.e. P2)
QAR	Quick Access Recorders
RA	Radio Altimeter
RAT	Ram Air Turbine
RPV	Remotely Piloted Vehicles
SAR	Search and Rescue
SAS	Special Air Service (UK Military)
SHELL	Software, Hardware, Environment, Liveware, Liveware (model)
SMS	Safety Management System
UAV	Unmanned Aerial Vehicle
ULB	Underwater Locator Beacons
UTC	Coordinated Universal Time (successor to GMT)
VHF	Very High Frequency (radio telephony)
VOR	VHF Omni Range (beacon)

1

Introduction

Regardless of where a passenger may travel from and to, travelling via commercial airlines is a safe and secure means to allow us to get from A to B in a relatively short time frame. Commercial aircraft travel is very different from surface transportation options, such as travelling by train, coach/bus/car or even a ship.

Yet commercial aviation has always been operated very differently from other transportation types. The security aspects (at the airport) are very different for travellers, staff and operators compared to the surface transportation means. But some of the biggest successes from changes and improvements have been driven by the aviation sector's willingness to 'learn' from its errors. If we consider an Airbus A350, full of fuel, cargo and passengers on a long haul flight, the A350's take-off run will involve it speeding down the runway at velocities greater than 155 knots (kts), which is approximately 180 miles per hour or 290 kilometres per hour. The speeds during landing with full flaps reach similar numbers. The purpose of bringing these values to the attention of the reader are to demonstrate the high speeds associated with take-off and landing, and assuming the aircraft (i.e. A350-1000 MTOM) is close to its maximum take-off mass, it will be weighing is 310,000 kg. Put into context, aircraft have always been vehicles that are very lightweight (and strong) but 300 tonnes of mass travelling at 180 mph is some achievement, and risk. The aircraft is not only a feat of achievement in terms of design but also maintenance, and yet the ways the airlines and Air Traffic Controllers (ATC) control and operate these commercial aircraft has barely changed since the significant growth of aviation after the Second World War. When problems have occurred during the take-off run, the flight or the approach/landing phase, this combination of high speeds and lightweight structures has been aviation's 'Achilles heel' in terms of the accident rate, survivability and so forth. With those basic facts established, the aviation business has always carefully reviewed past experiences, evaluated how things 'went wrong' in forensic detail and made the necessary changes to prevent the same event from happening again.

The governments around the world very carefully regulate the whole aviation sector. The governments use the form of a National Aviation Authority (NAA) of a given country (e.g. Federal Aviation Administration, USA, Civil Aviation Authority, UK, etc.). These basic rules and methods to organise and regulate air transportation date back to the Chicago Convention,

December 1944, where 52 separate countries or states agreed how aircraft, airspace and specific events would be controlled and organised. The convention signing later leads to the formation of the specialist United Nations organisation, the International Civil Aviation Organisation (ICAO). In general terms, the NAA has the responsibility for the airlines that are operating from their country, the design of any an aircraft originating from this country, the licensing and examination of pilots, engineers, ATC, etc....

Thus, as aircraft have developed, so have the rules and regulations that surround all aspects of them: these rules are procedures evolving at a similar rate to that of the aircraft, with the overall objective to improve safety. For the most part, this professional attitude of openness, review and continuous improvement has made commercial aviation the safest means of transportation in Europe and North America, which the travelling public has taken for granted in recent years.

The standards that are currently imposed for travellers regarding aircraft cabin security have evolved over the years, often as a result of hostile, and sometimes homicidal acts. While the security aspects for commercial aviation and travellers are so different from other surface transportation means, the justification for the higher levels of screening/profiling and checks is based on the historical events.

This book will explore the various aspects of commercial aviation's safety and security to evaluate significant past experiences (that are mostly in the public domain), to explain how some of the technical systems function in-flight, certain events that have resulted in the loss of the aircraft (due to deliberate acts) and lastly to explore how new state of the art technologies can be used to improve the high levels of safety that we, the travelling public, all take for granted. By understanding these aspects, one can better understand how the industry has evolved with successive events, the various incremental steps taken to stop 'hostile acts' from occurring inside a commercial aircraft.

Note: This book will not address (in forensic detail) the 'landside to airside' security screening processes carried out by security staff at airports. While the evolution of the screening is briefly explained, the technologies behind the latest full body (millimetre wave back scatter) passenger imaging (see Figure 1.1) or the latest conveyor based 'X-ray' scanner for bags/coats are not included. This is because the technologies underpinning these developments are rapidly changing, in terms of the computing algorithms that evaluate the sampled data. These predict whether the liquid in the bag is nail varnish remover (a formulation containing acetone), or a deadly liquid cocktail containing concentrated hydrogen peroxide that would lead to a detection signal and alert staff to conduct further testing. Furthermore, the current capabilities of the 'landside to airside' security screening measures are so rigorous, combined with the extensive use of computing calculations to perform the repetitive tasks to provide an accurate detection, the probability of

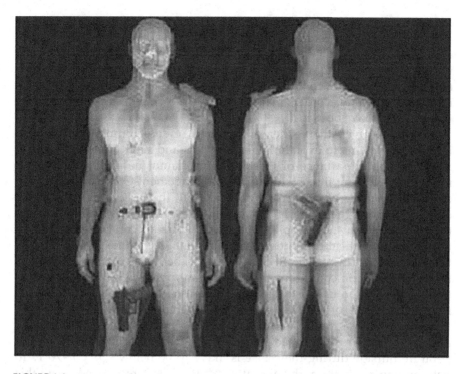

FIGURE 1.1
Example of a modern full-body (millimetre wave back scatter) image showing hidden weapons concealed in clothing. (L Kaufman.)

an individual being able to bring a prohibited item onto an aircraft is now very small indeed. The human operators that previously needed to view the 'X-ray' screens were never able to maintain absolute focus, because fatigue and repetition affect the security operators' ability to analyse the silhouette images and identify prohibited items. As much of the travelling passengers are also recorded during this in-depth screening, the potential that a traveller could avoid/obstruct or interfere with the security screening process and then be able to board a commercial flight is considered even less probable: this is because, should a failure in the screening process be detected (and be captured from the video data), it is possible (and for certain flights normal practice) for the 'airside passengers' to be rescreened for a second time, to establish an additional layer of certainty that all the passengers are free from prohibited articles (sharp objects, guns, etc.).

Chapter 2 discusses some of the more pertinent events associated with security aspects and failings in aviation. The context of why aircraft have been targeted over the years by successive hostile persons (insurgents/ terrorists and criminals) is explained, including the first recorded pilot homicide event.

Chapter 3 explains the background associated with Flight Data Recorders and Cockpit Voice Recorders. While these devices have their roots firmly established in early aviation, the use of wide spread data recorders became popular in response to unexplained accidents. The development of recorders is explained, from continuous magnetic media-based devices to today's modern solid-state recorders that have proved so vital in explaining why so many aircraft have been lost.

Chapter 4 gives the reader a technical understanding of two very important systems to explain Flight Controls and Environmental Control Systems. The importance of these systems cannot be underrated, as all modern aircraft are fitted with a Flight Management Computer (FMC) to define the routing of the flight. After take-off, the FMC uses the various data inputs from the aircrafts' navigation systems to fly the aircraft as per the routing when the autopilot is engaged. The air-conditioning system forms are the critical components that are included in the Environmental Control System (ECS). The ECS is necessary, because commercial aircraft fly in atmospheric conditions that do not support life in ambient conditions. The solution used is to take clean high-temperature compressed air from the engines (or for B787 an electric supercharger motor) and to pass this hot air into a device known as an air-conditioning pack. The pack allows for hot air to be piped into the aircraft in-flight, because the temperature outside during the cruise is likely to be below $-57°C/-70°F$. To help human respiration, the aircrafts' cabin is also pressurised using this continuous hot air flow, because atmospheric pressure (around 1 atmosphere/ 101,325 Pa/14.7 psi) is required to allow our lungs to exchange oxygen gas for carbon dioxide gas. If the cabin pressure falls, then the gas exchange in the lungs drops resulting in incapacitation (hypoxia), and if not remedied, death. The level of pressurisation is controlled by the use of two or three gas outflow valves fitted into the skin of the aircraft. If the aircraft senses that the pressure inside the cabin is too high, the computer controller opens the valves a little to release some extra pressure. The use of the ECS is explained in the same level of detail given to pilots and engineers, because this life support has implications in certain security events that will be discussed in Chapter 8.

Chapter 5 introduces how the aircraft maintenance management philosophy has developed. The overall approved maintenance planning is explained, the differences between the aircrafts' scheduled and unscheduled maintenance activities and the legal implications for these unscheduled maintenance defects that are deferred by the engineering operation to be repaired or resolved at a later time. Reliability in general terms is explained, because of the changes in the maintenance philosophy that have evolved since the Second World War. The financial gains that are attributed to an effective maintenance strategy to an airline are presented, because if an aircraft cannot fly due to technical reasons, this will cause an operator to lose revenue – thus the management strategy has very significant financial implications for the airlines.

The sources of the aircraft systems' performance data are explained, including how the old multiple-page carbon paper-based systems that were popular for many years were superseded by more modern technology. The real-time delays encountered between the engineering technical paper log being filled in and the time when the data is entered into a maintenance IT system are explored. New sources of data are included, including where these new sources are located on the aircraft. The methods for the engineering departments to use engineers with portable data storage devices that can interface with the aircraft's digital flight data recorder to download the data and subsequently be uploaded to a maintenance planning server are included. Likewise, being able to review some 'live' data from an aircraft's system when the plane is operating is included, following the introduction of the Aircraft Communication Addressing and Reporting System (ACARS). Other more up to date systems are cited, including the use of a ground-based Wi-Fi network fitted at the airlines home-base (i.e. *brouters*) or the engine manufacturer's inclusion of a data communication module (3G mobile phone based) on the engine's Full Authority Digital Engine Control (FADEC) module, to transmit the leased engine's performance trends (from the FADEC's data storage) back to the engine manufacturer.

The theory behind some of the predictive maintenance is included, to give the reader an appreciation that a 'learning mathematical system' is used by the likes of Rolls Royce to predict acceptable or non-acceptable performances, and how the software optimises to maintenance, i.e. when to change components or whole engine modules.

Major aircraft manufacturers have also worked closely with data analytic organisations to provide a subscription-based maintenance management service: Airbus's Skywise is used as an example, yet the data from the DFDRs still require an engineer with a laptop to download the terrabytes of data (from the week's operation) that need to be uploaded to a server.

The general theme and driving principle for all maintenance management systems is to optimise maintenance activities, to keep the aircraft flying for longer and to estimate from historic data when to change components. The financial gains using these technologies are very significant!

Chapter 6 explores the evolution of the aviation sectors' ability to learn from its mistakes (from historical events). Previous mistakes by pilots were simply explained as 'pilot error' yet the recurring historical crashes due to such pilot errors did not adequately explain why these highly trained professional pilots were behaving this way. After the 1972 Staines crash (a town very close to London Heathrow Airport), an accident investigator used his knowledge of the event and conceived the 'SHEL model' which explained the interactions between the pilot(s) and four other factors. The model was later surpassed by the 'Error chain' and the differences between these two models are explored. The third more recent human error model, known as the 'Swiss Cheese model', is explained, and the explanation as to why this model is so important to explain (Human Factors) deviations from intended performance is presented. Lastly, the overall Safety Management Systems of airlines discusses the legal requirements, and how these systems required the upper management to produce documentation specifying their corporate culpability should a significant failure occur. The general theme of the Human Factors concept is employees do not usually carry out reckless acts, rather they go to work to be productive, to do a role as best they can. However, deliberate acts, such as sabotage, are not errors but chosen decisions, and therefore must not be considered in the same context as errors.

Chapter 7 explores the physical barriers that separate the passengers from the pilots on commercial aircraft. The evolution from small sized aircraft with curtains between the pilots and passengers to bigger commercial aircraft with cabin bulkheads and flight deck doors are explored. However, the changes these 'barriers' make alter the way pilots and cabin crews work, as evidenced by the British Midland disaster in 1989. The reader will need to appreciate the importance of Human Factors from the previous chapter, to understand the need for open and effective communications between all the crews. The 'downside' of having a flight deck with an unlocked door is discussed, using the British Airways London Gatwick to Nairobi, Kenya event in December 2000 as an example of the dangers of passengers gaining unauthorised access to the flight deck.

The security implications of 2001 are presented, including the world's biggest single insurance loss of multiple aircraft, caused by the Sri Lankan civil war. The suicide attack on Bandaranaike Airport in July 2001 resulted in multiple military aircraft being destroyed (in the military's area), with the insurgents later attacking the civil airport on the other side of the airfield.

The infamous events of 9/11 are considered, including the immediate changes to the way the airport security operated. Restrictions on the passenger's 'carry-on' items changed, and the widespread deployment of Sky Marshals is was required. The manufacturers modified the aircraft's flight deck door, to prevent passengers from gaining unlawful entry into the flight deck when the aircraft was operating.

Further passenger airside security restrictions are discussed (re. carry on liquids) and the background behind the security-based decision to limit the volume of fluids. The security services' counterterrorism intelligence data was the trigger for the cabin liquid's restriction, and it was able to disrupt numerous attacks on UK–USA commercial flights.

Chapter 8 explores several unusual losses of commercial aircraft between 1983 and August 2018. The case studies use the open data from the accident investigation reports that are published by the Accident Investigation branches from their respective territories. The case studies are not intended to replace formal accident reports, because each subsequent report is several hundred pages long in addition to the appendices. Rather, some of the identified factors are presented, allowing the reader to understand the context and find the similarities across all the case studies shown.

The case studies include the Korean Airlines B747 shot down by Soviet forces (and references to other shoot downs), where the flight crew failed to set the navigation system correctly when leaving Alaskan airspace. The Pacific South West airlines homicide/suicide event (including crash) demonstrates the ability for employees (and ex-employees) to use their expert knowledge to counter the security protocols and execute multiple homicides (using a revolver) before crashing the aircraft, killing all occupants. Included is the EgyptAir New York to Cairo homicide/suicide and ditching, highlighting the significant differences in the accident report from both the USA and the Egyptian reports, and the determination of one of the First Officers to carry out this dastardly and deliberate act, even when the Captain returned to the flight deck. The SilkAir B737 crash cast study is discussed, and the publication of the accident report cited that the investigators believed the occurrence was a homicide/suicide crash. An unusual element was immediately prior to the near-vertical dive towards the ground, and coupled with the findings that the DFDR and the DCVR had both stopped recording in the minutes before this crash. These systems would require human intervention to switch them off: a later civil case in the USA courts was brought by relatives of several passengers, suing the manufacturer of the rudder controls, although the court refused to allow the accident investigators to give their evidence. The LAM Mozambiqe airlines pilot homicide/suicide/crash case study further highlights the dangers of pilots with underlying problems at home, including financial hardship, depression from earlier losses, etc. This case study highlights the problems when the second pilot leaves the flight deck (to a single pilot) to use the bathroom, and the mentally impaired pilot locks the flight deck door with tragic consequences for the aircraft and occupants.

The Malaysian MH370 disappearance case study is presented in detail, including some additional theories that have been published in late 2019 further to the official Malaysian accident investigation. The case study will suggest a hypothesis to explain the disappearance of this aircraft, the implication of the ECS and pressurisation systems (including altitude exceeding the aircraft's official safe service ceiling), the Malaysian investigations report findings re. the potential motives of the Captain, the use of the Captain's home flight simulator, and lastly the admission by the Malaysian authorities that the Boeing Uninterruptible Autoland system could not have been fitted to this aircraft.

The German Wings case study from the March 2015 event highlights the real dangers of mental health in active pilots, and the lengths that some of these persons (i.e. the first officer) will go to hide their rapidly deteriorating mental health problems, resulting in a homicide/suicide/deliberate crash. Additionally, the case study also highlights how the medical professionals that treated the pilot prior to the final flight did not report their professional findings to the German government's pilot medical unit (with the LBA), due to concerns of privacy for the individual. This case study has a similar theme to the LAM Mozambique event!

Lastly, the HorizonAir theft/suicide case study is slightly different as it was not investigated by the accident authorities, but rather the Federal Bureau of Investigation (FBI). The case study highlights how a ramp agent employee was able during his shift to enter an aircraft, start both the engines and take-off without permission – in short:theft. The employee that committed this criminal act went on to fly erratically after taking off (having never received any flight training), conducted very low level flight manoeuvres that were filmed by passers-by. The aircraft was intercepted by US Airforce F-15 aircraft and rather than landing and facing justice for this criminal act, the sole occupant decided to commit suicide by crashing the aircraft into an unpopulated island. The lack of security within a commercial airport is apparent, as there are no physical keys or codes used to start multi-million dollar aircraft – only switches.

Chapter 9 draws the findings of the earlier chapters together, to hypothesise the technologies necessary for minimising losses. The technologies associated with remotely piloted aircraft are tracked back to the First World War, but the developments made were kept secret for many years to come. Further significant developments in remote-controlled aircraft were made during the Second World War, and by the end of the war the US military had flown a large cargo aircraft from Hawaii to California, with the pilots flying the aircraft from the landing airfield. The development continued post-war by the militaries, often using the modified operational aircraft to become 'drones' that were used in shoot down target practice.

An unusual, unmanned series of flight tests were carried out by the Federal Aviation Administration (FAAs) full-scale crash tests using a Boeing B720 four-engine jet. The preparation of the crash test demonstrated that the aircraft could take-off and land, while being fully controlled from the ground. The Gulf War (Operation Desert Storm) in 1991 was the turning point for the US military's show of technological superiority in press briefings, as they shared images taken by these unarmed drones to explain US military achievements. However, some of the footage also included the enemy retreating/regrouping, yet as the drone was unarmed, there was little the drone operator could do. After the Gulf War, newer drones were developed, some carrying Hellfire missiles, others carrying surveillance equipment, and these drones had a combination of autonomous and ground-controlled operations.

In 2003 the Boeing Company submitted a patent claim for the Uninterruptible landing system, including reference to possible problems with pilots flying aircraft behind locked flight deck doors. Later during the year the Airbus group, including Honeywell Electronics and BAe Systems, made the astonishing media announcement that they had also been working on an automatic landing system, and that they had almost competed development. In light of these public admissions by both major airframe manufacturers, how these systems might function is discussed, including reference to additional 'micro-sized' chip-based navigation systems and integration into the modern instrument landing system. The technology to take over an aircraft experiencing a security event is also proven, with Airbus's fully Autonomous Taxi, Take-off and Landing (ATTOL) project, including video footage of the working system aboard the Airbus A350 aircraft that was completed in Spring 2020.

The chapter also hypothesises what might happen if a hostile occupant that is within the flight deck realises that the aircraft is poised to land because the manufacturer has activated an Uninterruptable landing system – will they simply sit back and wait for the inevitable?

Further current aircraft weaknesses are examined, and a series of modifications suggested to be undertaken to prevent further harm/loss of life, because the aircraft remains potentially vulnerable until it lands and is met by the armed security services.

Other additional modifications are proposed to counter the possibility of a repeat of an MH370-type event. For example, using the satellite data broadcast to continuously transmit (compressed packets) of data to the ground server. Other changes include preventing pilots from switching off the transponder in-flight (because it is unjustified) or being able to disable other critical systems, such as the ECS during a flight. Single-use data codes need to be employed in the industry to allow authorised users (i.e. pilots) to start the engines and fly to the given sector – and deviation will be immediately flagged and followed up by ATC. Lastly, some of the system's electrical circuit breakers should be moved from their current position, and the justification is presented.

Finally, the justification to make all such modifications will not be performed on the grounds of safety, because of the public's expectations coupled with the incredibly low event occurrence rate.

The driver for any changes must be based on the potential financial savings of a better/leaner/'greener' operation of commercial aircraft.

The cost savings will only be possible with the use of much more highly detailed flight data transmission from all the aircraft systems (including some video footage), all in real-time via satellites. The inability to steal or deliberately crash a large commercial aircraft will be a welcome by-product of the new technologies employed.

2

History of Security Events

2.1 Introduction

Aircraft, by their inherent high-value nature, are considered major assets. Their conspicuous value is evident with their premium rate engine technologies and their ability to transcend humans' abilities to travel; aviation has made it possible to be half-way around the world in the matter of hours, whereas previously it may have taken weeks or months. This results in aircraft being a significant mode of travel in an increasingly interconnected and globalised world. It is not only the high cost of manufacturing and maintaining aircraft that make aviation a significant target of hostilities, but their subsequent highly political nature. Aircraft disruptions have major ramifications that are distinct from other surface transportation means, due to a zero-tolerance attitude regarding failure. Consequently, this results in aircraft being able to consistently dictate high publicity in the media. Thus, as a result of these variables, aircraft have been hijacked to exploit these distinctive features.

Note: Explosive events, such as bombing of aircraft, are outside the scope of the discussion, and are thus not included. This is because the perpetrator of the event typically did not elect to fly with the device.

2.2 Threat and the Use of Violence in Aviation

The use of or threat of the use of violence has been frequently utilised in the history of aviation, dating back to the early days of flying. Historically, this has often included the use of weapons or explosives on board. The threat of violence, to gain a benefit through aircraft transportation, is commonly referred to as 'hijack' events. There have been records regarding the theft of aircraft; however, due to the essential and particular skill levels that are required to operate an aircraft, such occurrences are very rare in comparison to hijackings.

The earliest recorded event of a hijacking was on 21 February 1931, relating to a Ford Tri-motor aircraft (shown as Figure 2.1) located in Arequipa, Peru. After landing the pilot, Byron Richards, was approached by a local revolutionary militia Richards was detained for 10 days, as the militia attempted to force Richards to fly to another destination, to which Richards refused. Eventually, after notifying Richards that the revolution was successful, he was freed on the condition that he flew one of the members into Lima. This event was the first detailed account of a new form of hijacking: terrorism.

On 25 September, 1932 - about a year later on the same South American continent, during the Brazilian Constitutionalist Revolution, there was another theft of an aircraft, which included the kidnap of an airport staff member. Three rebels had taken a Panair de Brazil, Sikorsky S38 aircraft from a hanger. The S38 was a multi-engine piston amphibious aircraft (illustrated in Figure 2.2), seating around eight occupants including the pilot. While the perpetrators appeared to have no formal flying experience, they were able to take-off. However, they experienced difficulties controlling the plane and

FIGURE 2.1
A historic Ford Tri Motor Aircraft in US company livery. (Joe Osciak.)

FIGURE 2.2
A Sikorski S38 aircraft (replica). (Christian Bramkamp.)

unfortunately the aircraft crashed in the Sao Joao de Meriti area, killing all four occupants.

This incident highlights that while there are difficulties in acquiring an aircraft by force, the complexities in the preparation of the aircraft and necessary skills pilots must possess in order to take-off, navigate and land are critical. The consequences the lack of expertise are often loss of human life for both victims and those conducting the hijacking.

2.3 Murder at the Controls of an Aircraft

The first recorded murder on a plane was on 27 October 1939, committed by trainee pilot Earnest Pletch. While undertaking flight instruction in a Taylor Cub two-seat single-engine piston aircraft (see Figure 2.3), Pletch shot his flight instructor twice in the head, before landing the aircraft. This event was also the first hijacking in the USA and is considered very unusual due to the lack of a clear motive. In court, Pletch was found guilty of the act of homicide and was sentenced to life without the possibility of parole. However, Pletch was later released in 1957, having been incarcerated for 17 years.

FIGURE 2.3
A Taylor Cub training aircraft. (Max Nüstedt.)

2.4 The Second World War, and the Immediate Period Thereafter

During the Second World War, hijack events were rarely recorded, but a small number of events did transpire. After the war, with the repurposing of old military aircraft for civilian use, aviation operations flourished worldwide. As a result of the expansion of commercial air travel, hijackings exponentially increased and consequentially, they began to command worldwide attention.

Like many other commercial aircraft of the era, the Catalina 'Miss Macao' seaplane (see Figure 2.4) was a repurposed military aircraft.

The Cathay Pacific Airways Catalina aircraft was chartered by the subsidiary Macau Air Transport Company and was normally used to carry passengers. Being a seaplane, the Catalina was different to the other fleet of operational aircraft from the owners, Cathay Pacific Airways, as Cathay's first aircraft was a DC3 known as 'Betsy' (see Figure 2.5).

However, the Macau Air Transport Company also imported gold from Hong Kong to Macau, which would then be distributed to anti-Communist organisations who required funds, thus avoiding a violation of the Bretton Woods Agreement, which included the prohibition of the importing free

FIGURE 2.4
'Miss Macao' Catalina aircraft.

FIGURE 2.5
Cathay Pacific Airways first aircraft, DC3 'Betsy.' (Cathay Pacific Airways.)

gold among countries such as those in Western Europe (Hong Kong was a colony of the British Empire at the time). As Macau was a Portuguese colony in the Southern China region, Macao Air was exempt from these limitations (i.e. not covered by the Bretton Woods Agreement). While the 'Miss Macao' represented a growth in profits for Cathay Pacific and was an example of the global expansion of the aviation industry, it was operational during a postwar society where the security risk and potential for hijacking events were greater than ever. On 16 July 1948, a few minutes after take-off, four men hijacked the 'Miss Macao' seaplane, 6 km North East off Kauchau Island, China; the aircraft was enroute from Macao to Hong Kong. The hijackers had handguns and instructed the pilot to land in a remote area of the coastline in order to rob the passengers and ransom them. It was later reported that four of the passengers were significantly wealthy individuals, with one carrying over 100 kg of gold. Following an altercation with the hijackers, the pilot was shot in the head and collapsed over the controls, which resulted in the aircraft nosediving into the sea. All passengers and crew died, except for one hijacker who was the sole survivor.

While this was not the first hijacking of a passenger aircraft, it was the notable milestone involving a significant loss of life that resulted in protocol change to protect lives in such unforeseen events. Following this event in Asia, airlines around the world adopted a compliance stance with hijack demands, whereby they would comply with the demands of the hostiles to minimise the risk to life and to the aircraft. This also demonstrates the beginning of aircraft security evolving in response to hijacking incidents.

2.5 The Late 1950s and 1960s

In the years prior to the Cuban revolution (1958), a number of aircraft hijackings took place in Cuba, where a series of events occurred. This is where individuals (sometimes travelling as passengers) used force to take control of the flights, including the hijack of a Cubana de Aviación DC3 aircraft (Figure 2.6).

The hijackers' common motive for many of these hijack events was to escape the island of Cuba, prior to and in the years after the revolution for independence. Between 9 April 1958 and 2 October 1959, there were at least 11 separate recorded events. The occurrences of persons using force to hijack flights bound to or from Cuba became even more frequent in the following decade. The solution that the United States Federal Aviation Administration (FAA) adopted was the deployment of armed guards on commercial flights from 1968 onwards. The introduction of these armed 'covert' guards (later known as 'Sky Marshals') did little to prevent the regular occurrences of hijack events on flights between the USA and Cuba (and vice versa).

FIGURE 2.6
Cubana de Aviación DC3 aircraft.

The American Airlines further responded to these frequent hijack events (that were very bad publicity for the carrier) with an 'absolute compliance' to the hijackers demands. Pilots and crews would obey the instructions given by the hijackers.

2.6 The 1970s Onwards

Throughout the 1970s numerous further acts of terrorism took place on USA–Cuba flights. The regular occurrences took place up to late 1972. In 1968, Senator George Smathers (Florida State) suggested that 'new' security technologies that were being used at US high-security prisons should be deployed at US Airports to prevent the hijacking problem. Shortly after this suggestion by Senator Smathers, the FAA dismissed his concept. However, the government and industry later deemed the concept as a viable solution to the problem of aircraft security, as July 1970 saw New Orleans International Airport becoming the first airport to employ magnetic detectors as a means to detect weaponry. Passenger profiling was deployed as the initial means used to identify 'who to search', which might result in a further 'pat down' search of the traveller. Later, all passengers were required to walk through a

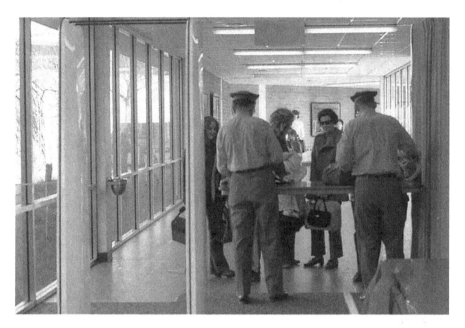

FIGURE 2.7
Metal detectors used to enhance security inside airports during the 1970s.

'hoop' style metal detector (Figure 2.7) when proceeding from the landside to the airside.

The 1974 US Air Transport Act saw all airports being required to introduce metal detectors for all passengers, and X-ray machines to screen carry-on baggage (Figure 2.8).

Some of these hijack demands during the 1970s included the ransom of the aircraft, where the threat of force was implied either by use of a weapon or by the suggestion that the person was carrying a bomb. Payments were demanded by the perpetrators and paid by the airlines to secure the release of the aircraft and crews.

Elsewhere in the world, hijack events also continued on a regular basis, often with a deeper political motive. For example, on 31 March 1970 Japan Airlines Flight 351 on a Tokyo to Fukuoka domestic flight was hijacked by the Japanese Red Army. The flight was forced to land in Gimpo Airport, South Korea. After a standoff, the hijackers demanded a further onward flight to Pyongyang, North Korea. Likewise, 6 September 1970 saw a politically motivated terrorist group (The Popular Front for the Liberation of Palestine) coordinate multiple hijackings of four separate aircraft (TWA flight from Frankfurt, Swissair from Zurich, *attempted hijack of El Al from Amsterdam*, Pan Am from Amsterdam, BOAC from Bahrain). All the hijacked aircraft were forced to fly to Jordan, and land at a remote site known as Dawson's Field.

FIGURE 2.8
X-ray baggage scanners used at airports and high-profile security events by US Secret Service during the 1970s. (American Crystallographic Association, David Haas.)

As the years continued from the 1970s to the 1980s, the increased levels of passenger security screening did prevent many of the 'would be' attempts to hijack aircraft and make demands. While the number of occurrences in the USA continued to fall year on year, as a result of the more effective screening, yet occurrences elsewhere in the world continued frequently.

Security and screening continued to evolve in the 1980s and 1990s, and the carrying of prohibited items by passengers onto large commercial aircraft became more difficult, with requirements to pass through landside/airside security thwarting many attempts. However, some hijackers changed their methodology to use some of the emergency equipment carried onboard the aircraft. One example of this was Ethiopian Airlines Flight 961, which was a multi-sector flight from Addis Ababa to Lagos in Nigeria. After take-off, three young men stormed the flight deck, obtained the fire crash axe that is carried on aircraft for fire-fighting purposes, and threatened violence to the pilots. The fire crash axe is carried to pry open panels and bulkheads to gain access a fire in an inaccessible location. One of the hijackers on this

FIGURE 2.9
Hijack and ditching of Ethiopian Airlines Flight 961 on the beaches of the Comoro Islands.

flight falsely claimed he was also carrying a bomb onboard the aircraft, and the three terrorists demanded to be flown to Australia. The Captain of the aircraft knew that this B767 aircraft was not carrying sufficient fuel for the demanded flight, and instead flew down the African coastline in a general southern direction. The Captain directed the aircraft towards the Comoro Islands, but before the aircraft could reach the airport and land, the aircraft exhausted its supply of fuel. The aircraft then performed a forced (deadstick) ditching on the sea, some 100 m from the beaches that were popular with tourists. Tourists on the beach were able to film the B767 ditching, cartwheeling in the process (see Figure 2.9), before sinking to the seabed. A local dive school returning to the beach on boats, having just finished a series of underwater diver training classes, was able to respond immediately. The members of the dive school were able to descend onto the submerged B767 that had sunk in shallow water: the divers rescued 50 of the 175 passengers and crews.

The changes made after the hijack events associated with 9/11 will be discussed in Chapter 7.

2.7 Conclusions

This chapter has explained a brief history of some of the hijack events where perpetrators have used or threatened to use violence to coerce the airlines and the pilots to submit to their will. The historical events indicate that hijack is not a new problem, because aircraft and aviation is seen as a high-value target. Early demands were to force the pilot to carry the hijackers, but this later changed to extortion and politically motivated acts. Pilot homicide inflight was also observed. Eventually, the regulators and industry responded to this new threat of hijacking, with the introduction of Sky Marshals on certain flight routes, and improved security profiling of passengers. The mandatory use of 'hoop' metal detectors and cabin baggage X-ray screening for passengers moving from the landside to airside did result in a decrease in the event frequency, but many individuals were still able to circumnavigate these new security protocols, as the hijack statistics demonstrate. Lastly, one hijack event included storming the flight deck and obtaining the aircraft safety equipment fire axe to coerce the flight.

The next chapter will explain the background events behind the development and deployment of flight data recorders and cockpit voice recorders (i.e. black boxes), which are used by accident investigators to piece together the final moments before a crash or ditching event.

3

Flight Data Recorders and Cockpit Voice Recorders

3.1 Introduction

Flight Data Recorders (FDRs) and Cockpit Voice Recorders (CVRs) were introduced as a tool to aid accident investigation after a serious and often catastrophic complex event. The need for such devices arose because traditionally, aviation failure events did not provide many 'first-hand witnesses' to such failures, especially when it involved aircraft flying at high altitudes. An example of the need for recording devices stemmed from the 1950s de Havilland DH106 Comet, the world's first commercial aircraft (Figure 3.1), and their successive unexplained crashes.

This type of aircraft entered commercial service with the British Overseas Airways Corporation (BOAC) on 2 May 1952. While a small number of unforeseen events took place in 1952 and 1953, the losses stemming from the successive disasters of 1954 are what made this aircraft infamous. A series of aircraft suffered catastrophic losses in-flight, and at the time it was noted that the loss of the aircraft the absence of any radio communications. The accident investigations were inconclusive, and the crash investigation teams were unable to immediately determine the cause of the accidents. While some debris from these events was recovered, the principal causal factor was not identified at the time that the debris was recovered.

The manufacturer (de Havilland) conducted a static pressurisation water tank testing (Figure 3.2) on aircraft G-ALYU. After a number of simulated pressure cycles were reached, the test airframe suffered critical failure, bursting around riveted joints (Figure 3.3) which in turn failed the mechanical resilience of the fuselage. This mechanical test was the pivotal investigation technique that would shape the industry, identifying a knowledge gap on flying aircraft.

FIGURE 3.1
BOAC Comet 1 Aircraft.

FIGURE 3.2
de Havilland Comet G-ALYU in the water tank for pressure tests.

FIGURE 3.3
Square window failure of Comet G-ALYU post the water tank pressure tests.

It was only due to these static tests that were performed that the manufacturer and accident investigators were able to explain the previous accident's principal causal factor. Once the root cause of the failure was determined, the remaining flying aircraft were modified to remedy this in addition to future production aircraft.

3.2 Flight Instrument Recording

Early attempts at recording flight performance characteristics were performed in the 1930s by François Hussenot and Paul Beaudouin in France. The recorder used light-sensitive photographic film to capture images of the critical flight instruments (altimeter, airspeed indicator, etc.) at a pre-determined time interval. This is the earliest recorded instance of a device being used, instead of a flight crew member taking such readings manually.

After the losses of the Comet aircraft in 1954, David Warren, an Australian research scientist understood the difficulties in the accident investigation processes, especially for those that had no natural causation factor. His solution to this problem was to describe the theoretical concept of a data and voice recording device that would be installed on the aircraft, in a peer-reviewed paper: 'A Device for Assisting Investigation into Aircraft Accidents' (1954). While Hussenot's devices recording devices were further developed in the 1950s, the lack of information in the aircraft investigations was problematic. Warren had a personal interest in the field, having lost his father to an unexplained plane crash as a young child. Warren developed a working prototype in 1956, which later went into production utilising steel wire as a means to record the cockpit voice recordings. Warrens device was known in the industry as the 'Red Egg' due to the colour and shape of the device casing.

Further FDRs and cockpit sound recorders were developed in the 1960s by Prof. James Ryan and Edmund Boniface, respectively, in the USA. Both such devices were patented and introduction of these recorders was slow to gain popularity. One such obstacle was the flight crews themselves. They believed that the airlines would eavesdrop on their private conversations onboard the aircraft, compromising their ability to pilot the aircraft, effectively. This impasse was later resolved by flight crews and airlines, with the understanding that the recorded sound data would only be obtained in the event of a recorded event (e.g. a plane crash). The recorders typically operated by encoding the data within the unit onto a continuous magnetic tape/wire loop (Figure 3.4), with a continuous operation specification.

FIGURE 3.4
Flight data recorder fitted to a pressurised aircraft, exposing the magnetic recording wires.

FIGURE 3.5
An example of an early Flight data recorder fitted to Boeing 707.

The data recorders would continuously record up to 2 hours of flight performance data, whereas the CVRs record the last 30 minutes of audio. Early examples of such flight data recording (Figure 3.5) included just five parameters, e.g. altitude, magnetic heading, indicated airspeed, vertical speed indicator and the microphone – all recorded temporally.

The location of such FDR and CVR recording devices was situated in the tail of the aircraft, i.e. aft of the rear pressure bulkhead. The purpose of this location choice was based on the expectation that the tail would be most likely to remain intact post an event. The absence of jet fuel and minimised risk of fire was an important factor, as the recorders were both fragile and could be damaged in a post-crash fire event.

3.3 Certification and Flight Instrument Recording

Certification of commercial aircraft has traditionally been the legal responsibility of each sovereign nation that builds new aircraft. For example, for new aircraft certified in the United Kingdom, the Civil Aviation Authority has specified the minimum performance standards deemed necessary. The use of flight recording devices became mandatory in the early to mid-1960s by the respective regulators around the world. Such requirements specified which aircraft were to be included (judged by take-off mass, number of seats, etc.)

FIGURE 3.6
Orange painted Flight Data Recorder with distinct white stripes for recognition.

with remaining operational aircraft later included, with the need to retrofit these machines.

After a crash event, the recovery of the FDRs and CVRs was prioritised. To assist in this recovery, the cases of these devices were painted in bright, distinct colours to aid recognition (Figure 3.6).

Forced landing on water proved to be problematic for recovery by the search and rescue teams. The solution to this problem was the mandatory introduction of Underwater Locator Beacons (ULB) (Figure 3.7), which are battery-powered devices that emit an ultrasonic signature of a 10ms pulse once per second at 37.5 kHz ± 1 kHz for a minimum of 90 days. This frequency has been selected as there are no known natural sources that emit this output, and once the signal is detected, search teams are confident that the source is a ULB, allowing for a more effective recovery.

As recording technologies have matured and improved, new methods of storing data have become possible. Early data recording used coded information recorded as magnetic data onto a continuous metallic tape, and the limitations of the time meant that only a small number of parameters could be recorded. However, with the advancement of signal processing and storage technologies, the ongoing trend from 1985 onwards was to move away from magnetic tape storage (Figure 3.8), and to utilise solid-state recording.

FIGURE 3.7
Underwater Location Beacon fitted to FDRs and CVRs.

FIGURE 3.8
Magnetic tape-based FDR from a recovered aircraft post a crash event.

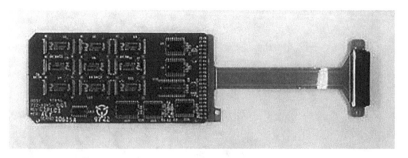

FIGURE 3.9
Solid-state-based data recording memory cards used in FDR.

The solid-state memory cards (Figure 3.9) permitted much more useful information to be recorded, thus the certifying manufacturing regulators (e.g. FAA, CAA, etc.) increased the number of monitored channels, frequency of monitoring and the total duration according to the technology 'of the day'. In later years, the agreed standards of recording have been published by the International Civil Aviation Organization ICAO, Annex 6 in the respective revision. These revisions state the minimum numbers of channels to be monitored and define the frequency of data logging necessary.

3.4 Decoding Flight Instrument Data from Data Recorders

Magnetic tape-based storage recording devices required a very specialist technical apparatus to download and interrogate the recovered information. These old FDRs were robust and designed to withstand the forces anticipated from a catastrophic event; removing the magnetic tapes was not a simple or straightforward task. The introduction of the solid-state recording memory cards (Figure 3.9) has allowed for more crash resilient FDRs and CVRs that are more likely to retain the integrity of the recorded data. Modern recorders have granted all the data to be recorded as discrete digital data values, rather than the older analogue data formats, which have been made possible by the improvements in data storage technology. As a direct result of this digital innovation, all data is stored in a digital format, namely Digital Flight Data Recorder (DFDR) and Digital Cockpit Voice Recorder (DCVR). The last recorded event of a crashed passenger aircraft using an FDR and CVR was on 24 November 1992 in China. A China Southern Airlines, B737-300 series aircraft crashed and while the CVR was not found, the FDR was recovered. Unfortunately, the FDR was found to be severely damaged in the post-crash

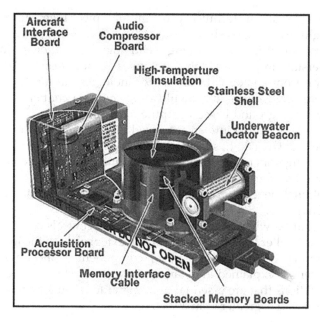

FIGURE 3.10
Solid-state-based Flight Data Recorders and Cockpit Voice Recorders including the underwater beacon.

fire exposing the tape to the environment. This resulted in the loss of flight data information.

DFDRs and DCVRs (Figure 3.10) are designed to withstand much more significant events than the earlier metallic recording based units, with the digital variants capable of retaining the recorded data without the loss or corruption of the information contained within the memory modules. This technological advancement has been possible by the new design of the units, with minimal moving components, the memory boards encased in a solid fire-retardant material that is in turn encased in a stainless steel shroud.

3.5 Conclusions

The introduction of FDR and CVR in large commercial aircraft has been driven by the need of the industry to identify the last moments of a flight before a crash. Early recorders used magnetic media that could only store a limited quantity of data, with any new data overwriting the previous content. The introduction of such devices was instrumental in helping accident investigators identify the technical problems and piece together the last minutes of a flight prior to a crash.

The improvement in the recording technologies has allowed for a much better understanding of accidents by investigation teams. Additionally, the transition from analogue to digital technologies has also permitted for a much more rapid means of downloading the data contained within the data modules to a ground-based computer. The flight performance data, for instance, can be uploaded into an approved flight simulator model, to recreate the same event in a flight simulator. This is particularly useful to allow for accident investigators to experience the same indications and characteristics that the respective flight crew observed.

The additional benefits of airlines utilising the DFDR data will be discussed in Chapter 5.

The next chapter will explain the primary flight controls of an aircraft, the automated controller (autopilot) that is used extensively in the industry. The air conditioning system, being a major component of the Environmental Control System, is explained in detail, because this system provides the survivable conditions that are essential to human life when an aircraft is flying at high altitudes (i.e. greater than 20,000 ft).

4

Flight Controls and Environmental Control Systems

4.1 Relevance of Flight Controls Air-Conditioning Systems and Commercial Aviation

In this chapter, it will be explained how pilots control the aircraft in-flight using a series of movable surfaces fitted to the airframe, the structure of the aircraft. The purpose of understanding this technical content is to recognise how these moving surfaces change the flight aerodynamics locally, the general design logic behind the control, and the most obvious differences between Airbus and Boeing (being the two most significant commercial aircraft manufacturers).

The pressurisation and ventilation systems on a large aircraft are considered to be essential to support life, and these systems will be explained in this chapter.

Both the development and technological advances of flight controls and the use of pressurisation systems have historical influences in the First World War and Second World War, especially in the later stages of the Second World War. As the wars progressed, the technology of the aircraft developed during both conflicts, respectively. The general trend was that the military aircraft became larger, flew further and carried more mass (i.e. bombs/fuel/ structure). Also, to improve their survivability in combat, these military aircraft flew at much higher altitudes to evade being shot down. By the end of the Second World War the largest technological advances were seen, with pilots and crews being pushed to the limit of their physical performance. Afterwards that journey of development continued, albeit at a slower rate.

4.2 Comparing the Underlying Philosophy of Flight Controls for Airbus and Boeing

Aircraft have been increasing in size ever since the first flight of the Wright Brothers Kittyhawk plane in 1903 (Figure 4.1). That initial flight was possible due to a combination of a gasoline motor coupled with human-controlled and 'powered' flight control technology. The wings were deformed in-flight by 'wing warping', which used cables to twist the wings to bank the aircraft. Yawing was achieved using two small vertical surfaces, again controlled by cables. The simplicity of the flight controls in the Wright Flyer allowed for those early achievements, and further enhancements followed thereafter.

As aircraft evolved throughout the two World Wars, a common theme that is apparent was the introduction of larger aircraft that could carry more mass, with an equally larger wing and control surface area.

Early aircraft were controlled with cable runs that were coupled between the controller (the yoke) and the moving surface (i.e. elevator, ailerons). Simple aircraft, such as single-engine piston planes such as the Cessna 172 or Piper PA28, still continue to use this technology, as illustrated in Figure 4.2. One such advantage for using this simple arrangement is the forces felt by

FIGURE 4.1
Wright Flyer, Kitty Hawk, first flight 1903.

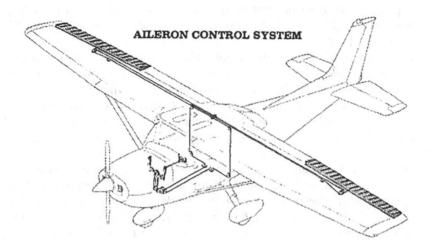

FIGURE 4.2
Cessna C172 simplified drawing of Aileron controls using wire cables coupled to the yoke. (FAA.)

the pilot when flying. If the aircraft increases speed significantly, the forces needed to move the surface increase exponentially: the pilot can feel with their hands the forces necessary to move the controls while flying the aircraft.

With much larger and heavier aircraft, the need arose to use significantly larger control surfaces with bigger wings. It became apparent to the aircraft designers that human pilots were approaching their physical limits. Bigger military pilots were starting to struggle to move the aircraft controls (i.e. the yoke), so new technologies were introduced post-Second World War to rectify this limitation. The cable-run technologies that had proved so popular earlier evolved to incorporate the introduction of hydraulically-moved surfaces. Cables were retained because the technology was mature, and more importantly they were much lighter than having long pipes full of hydraulic fluid that would otherwise need to run from the flight deck to the extremities of the aircraft. Weight is always a significant factor for aviation, as demonstrated by the military in both World Wars.

The use of hydraulics was a natural choice because the cables that connect the yoke ran to a valve block that was close to the moving surface. The moving surface was directly coupled to the hydraulic actuator, which in turn is controlled by the respective valve blocks. The introduction of hydraulic technologies resulted in pilots losing the ability to feel how the aircraft was flying, so a workaround solution was the introduction of 'Q-feel' – or artificial feedback into the flight control circuit. This Q-feel operated by measuring the airspeed of the plane, and mechanically making the input to the yoke more difficult at higher speeds. An additional development was to reduce the number of moving surfaces on the wings at high-speed flight compared to take off and landing, which reduces the wing loadings (and the forces acting upon them).

The incorporation of hydraulic actuators into flight controls has resulted in another technological development, namely the autopilot system. If an electrical or electronic control system is incorporated into the hydraulic valves (for the autopilot), the hydraulic actuator can directly control the movement of the surface. What this means is that the pilot's previous need to hold the yoke at all times is reduced, especially when the autopilot function is engaged.

Small corrections to an aircraft's moveable surfaces are made with trimming devices, and these trims are fitted to the elevator, ailerons and rudder, and operate independent of one another. Once again the trimming function was initially with cable-control, leading to hydraulic actuation, the same development process as with the main moving surfaces, including inputs from the autopilot.

All passenger flights on large commercial aircraft will require the pilots to make use of the Flight Management System, in addition to the navigation systems that are fitted on board the aircraft. When pilots prepare an aircraft for a journey, one of the tasks that will be undertaken is to programme the navigation computer with all the necessary details. During the flight, the aircraft calculates its current position from navigational inputs. These include Very High Frequency (VHF) radio signals, including VHF Omni-directional Range (VOR) signals; flight performance data, such as the altitude and airspeed from the Air Data Computer; magnetic compass readings; and inertia-based systems such as ring laser gyros.

The actual 'hands-on' control of the aircraft will be taxiing the aircraft from the stand (the terminal area) to the runway, the take-off roll and the initial climb. The usual commercial practice is to immediately engage the autopilot system in the post take-off climb, and to let the Flight Management System (the autopilot and the navigation System) fly the aircraft all the way to the destination. The controls for the autopilot are usually located on the glareshield, which is the central panel that is between the instrument displays and the front windshield. An example of this system is illustrated in Figure 4.3, representing the B747-400 aircraft.

The use of the autopilot system is a reliable, proven technology that, once activated, allows the pilots to reduce their own workload by using this system. The system can control all the axis of the flight, in addition to the speed of the aircraft and the rate of descent. The aircraft will fly the route towards the destination airfield, as per the pre-programmed plan. At around 10 nautical miles away from the airfield, the aircraft would usually be configured for a possible landing, meaning that a steady rate of descent (circa 3 degrees down) has been established. Usually, Air Traffic Control will be instructing the Pilots via radio commands as to what airspeed should be flown,

FIGURE 4.3
B747-400 Glareshield autopilot controls.

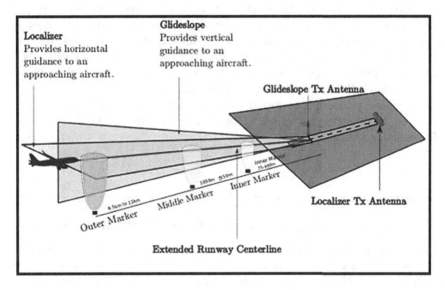

FIGURE 4.4
Aircraft approaching runway using the Instrument Landing System.

headings, etc., to ensure the safe operation of all the aircraft in this zone. It is at this time the aircraft's Instrument Landing System (ILS) has 'captured' the airport's precision landing radio signals, as illustrated in Figure 4.3 (vertical glide slope and the horizontal localiser radio signals). The use of the ILS as a precision landing tool is normal, as commercial aircraft fly in all weathers, and often the pilots cannot see the runway as they are approaching from a large horizontal distance. Before the landing, usually at a pre-agreed 'Decision Height' (which is based on the airports published Category class, such as Cat I, II, IIIa, IIIb or IIIc), the Pilot flying will make it known to their colleague their intention to continue to land, or to climb away and try again. Assuming that the pilots are continuing with the approach beyond the Decision Height, some modern aircraft may continue to be flown fully on this autopilot and Autoland system to the actual landing (i.e. the aircraft fully transitioning the mass to 'weight on wheels', if the airport and corresponding runway permits).

Lastly, the Airbus and Boeing flight control systems have very different underlying control philosophies. Up to the introduction of the Boeing 777 aircraft in the late 1990s, the Boeing flight control philosophy has always been to combine cables and hydraulics that are connected to the yokes, so if a total computing failure occurred on a Boeing aircraft, the pilots could land the plane manually.

Airbus has a very different philosophy. Namely, the flight deck does not have yokes, but two side sticks, like the devices used in computer games. The pilots move the sidesticks which change a voltage, and this change is interpreted by a flight-control computer. The Airbus flight computer flies the aircraft at all times, and the computer decides whether to allow the hydraulically-controlled surfaces to move in the direction that the pilot requests. If the Airbus pilot's sidestick movement is outside of the safe flight envelop during a flight, the pilot's sidestick input is ignored.

These two philosophies are drawn from the technical literature provided by the manufacturers, e.g. Flight Crew Operating Manuals, Aircraft Maintenance Manuals.

4.3 Air-Conditioning Systems and Commercial Aviation

A typical commercial aircraft has a maximum take-off mass greater than 5,700 kg and a cruising altitude that can range from 25,000 to 41,000 ft (Airbus 1992). At these altitudes, the outside temperature can fall to below −57°C and have a local outside air pressure of around 3 pounds per square inch (0.3 bar). To keep the passengers in relative comfort (and alive), it is critical to pressurise the aircraft to allow for effective human respiration, and to warm the passenger cabin. The air-conditioning system, which is also known as the Environmental Control System (ECS), is a **vital system** fitted to all large aircraft. Without an operational ECS at altitude, the occupants of an aircraft would suffer the effects of hypoxia and the extreme cold. Exposure to atmospheric conditions at high altitudes (c. 40,000 ft) in the absence of a functioning ECS will usually result in incapacitation after 30 seconds, leading to permanent incapacitation (death) thereafter (Airbus 1992).

The most important component within the ECS is the heat exchanger, and because of the weight and size constraints on aircraft, a plate-fin heat exchanger (PFHE) is typically employed. The air-conditioning system is often referred to in technical literature as the 'pack' (Airbus 1992). Hot compressed air is used by the aircraft ECS to provide fresh air for respiration; conditioned air to maintain the cabin temperature and; pressurised air to maintain the equivalent of an altitude of 8000 ft. The purpose of the ECS pack is to cool the hot bleed air from the engine compressors before the ventilation stream is passed into the passenger cabin in a large commercial aircraft. The ECS pack allows the flight crew to control the cabin temperature by adjusting the performance of the pack (Airbus 1992).

If the ECS fails in the air, the flight crew will attempt to land as soon as possible. If the aircraft is on the ground, the aircraft will be taken out of technical service and will require a significant maintenance action to correct the failure. It is not possible to continue to safely fly to the destination with an ECS failed or switched-off, because the passenger cabin will lose pressure quickly; the environmental temperature will fall to the same as the outside temperature; and the occupants will become incapacitated, as previously mentioned. Note: the use of emergency drop-down masks will not be effective in the event of ECS failure, because of the loss of pressurisation, which directly affects the human ability to exchange gasses (CO_2 for O_2) in the lungs, i.e. loss of partial pressure that results in asphyxia.

The air-conditioning systems that are fitted to modern large commercial aircraft are units that allow both the provision of cold and warm air on both the ground and while the aircraft is in-flight. Commercial aircraft are designed to operate worldwide in a variety of different environments. The common objective of aircraft designers when producing an aircraft is that the aircraft cabin environment should be capable of replicating 'comfortable' environmental conditions when the aircraft is 'in-service'. The term 'in-service' implies from the time the passengers embark into the aircraft, the period when the aircraft is taxiing for take-off, the take-off/cruise/descent phase of the flight, the landing and finally the final taxiing and disembarkation of passengers at the destination airfield. A comfortable passenger cabin typically is maintained at a temperature of 20°C, with a local equivalent 'cabin altitude' of no more than 7,000 ft.

The physical environment from the origin airfield via the flight phases to the arrival airfield changes significantly in terms of the temperature and local air pressure. In addition, the local conditions of the airfields at the departure and arrival points are likely to have different local environmental conditions, e.g. a flight departing from London Heathrow destined for Kuwait will have significantly different environmental conditions at the two airfields. For example, the local temperate conditions for London Heathrow are likely to require the cabin to be heated during the taxing phase within the UK if the outside air temperature on the ground is 10°C. After take-off, the outside air temperature decreases with altitude (lapse rate) until the aircraft reaches the tropopause layer (c. 37,000 ft) where the temperature at this altitude and above (20,000 ft) usually remains at a constant −56°C. The extreme cold conditions of the flight require significant levels of heating to be provided to the aircraft passenger cabin, to maintain the designers' required level of comfort. At the arrival airfield in Kuwait, the local conditions in summer are considerably hotter than London Heathrow, with the outside air temperatures typically reaching 50°C in the daytime. As the aircraft approaches the destination airfield and subsequently lands, the aircraft still needs to maintain the cabin environment of 20°C. While many airfields in hot climates do have the provision of ground air-conditioning equipment, it is worth noting that this service is only available for aircraft that are parked. Such equipment is useful for cooling the aircraft cabin prior to embarkation, but from the author's personal experience in Hong Kong Airport, once the external cooling is removed during the summer when the aircraft is in direct sunlight, the cabin temperature reaches 40°C in less than 30 minutes. Cooling is not solely a provision for passenger comfort, but more importantly, the air-conditioning system provides a cooling flow to the aircraft's electronics and electrical rack-mounted equipment. If the cooling air function is removed from the electronics and electrical bay, it is possible that the electronic systems will overheat and fail.

An operational consideration is that the aircraft typically closes the doors and begins ground manoeuvres some 30 minutes before the allocated take-off time, thus the need for the provision of cooling air to the aircraft.

The heating and cooling functions are required for a safe, comfortable passenger cabin environment on both the ground and in-flight, which are provided by the mechanical systems fitted within the aircraft structure. Older aircraft (pre-1960s) used a Freon-type refrigerant rig to provide the cooling function for the cabin air, but the weight, cost and poor reliability of the product made them obsolete in large commercial aviation after the 1970s. Due to space and weight limitations that are unique to commercial aviation, a novel system component was fitted to aircraft to utilise the engine's bleed air (high pressure and high temperature), to a device known as an air-conditioning bootstrap pack. The term 'pack' is used to describe all the mechanical components associated with the air-conditioning systems that are fitted to the aircraft. An aircraft pack is a mechanical device that uses a compressor and a turbine coupled with PFHEs to control the outlet air temperature and flow to the cabin.

Hot, compressed, clean air is taken from the gas turbine engines, and this is known as the 'bleed air'. The bleed air is ducted from the high-pressure compressors at a manifold, known as the customer bleed port. This hot air is metered through a bleed air valve and is required for commercial aviation to pressurise the passenger cabin, and to provide heating and fresh air for ventilation purposes.

The bleed air is too hot and too high pressure to directly duct into the aircraft cabin, as typically temperatures of the air are about 200°C. Hot bleed air from the engine compressors is metered through the bleed air valve, located in the engine pylon, on to the ECS pack: inside the pack unit. The air is further metered through a pack flow control valve (see Figure 4.5), passed through a primary heat exchanger (where there is a temperature reduction), then to

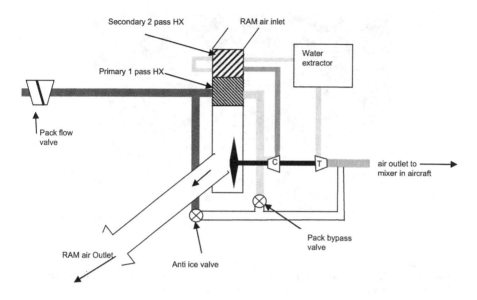

FIGURE 4.5
Schematic of the A320 air-conditioning bootstrap pack when the aircraft is on the ground.

the 'air cycle machine bootstrap turbo machinery' comprising of the compressor (C) and the turbine (T). The bootstrap compressor (C) centrifugally compresses the bleed air and thus raises the temperature and pressure, and the air is passed to a secondary heat exchanger, where there is a temperature reduction in the air. The air flows to a water extractor, which is required since ice crystals can form in the bootstrap which, if not removed, would cause significant damage to the turbine and associated turbo machinery. The air passes out of the turbine with an associated drop in temperature and pressure as the outlet bleed air expands rapidly. Finally, the cool, conditioned air is distributed into the cabin air system (Figure 4.5).

Ambient air, also known as ram air, provides the cooling in both the primary and secondary exchangers. The ram airflows are controlled by the electronic pack controller by opening and closing the ram air inlet flap and opening or closing mechanically louvered vents at the ram air outlet. When the aircraft is operating on the ground, i.e. while it is taxiing, the ram airflows are not sufficient to provide adequate cooling for the pack PFHE, even though both the ram air inlet flap and outlet louvers are fully open. The ram airflows are increased by the use of a mechanical shaft and fan which is coupled to the compressor/turbine shaft (Figure 4.5). As the bleed air is fed to the pack, the flow causes the compressor to rotate, the mechanically powered fan induces a ram airflow. The greater the bleed air supply, the faster the bootstrap turbo machinery and likewise, the greater the ram air cooling flow.

FIGURE 4.6
Photograph of the Airbus A320 on the ground, with the pack ram air inlets (fully open) and exhaust outlets, viewed from underneath the aircraft (facing aft.)

The ram air inlets are positioned below the water line of the aircraft, on the underside of the fuselage, located slightly aft of the wing root leading edge (Figure 4.6). When the aircraft is on the ground, the pack controller fully opens the ram air inlet flap to permit a maximum external cooling ram airflow to the pack PFHE.

Conversely, when the pack controller determines that full ram air cooling is not required, the ram air inlet flap is mechanically closed (Figure 4.7).

When the aircraft is in-flight, there is less need for providing significant quantities of cold fresh air to the cabin, and due to the cold external conditions as the altitude increases, warm air is required to heat the passenger cabin to provide a comfortable cabin environment.

The pack achieves the desired cabin conditions in a number of ways. Firstly, the ram air inlet flap is moved mechanically towards the closed position to restrict the cooling flow to the PFHE. The bleed airflow continues to pass through the primary exchanger, but by reducing the ram air cooling flow, the primary outlet temperature does not fall as significantly as on the ground. Additionally, the pack bypass valve (within the pack) is fully opened (Figure 4.8) thus allowing the majority of the bleed airflow to bypass the bootstrap turbo machinery. The use of the pack controller, which is a solid-state electronic control unit, allows either cold or hot air to be provided to the aircraft cabin, depending on the phase of the flight and the current physical conditions.

A simplified schematic of the full ECS (air-conditioning system) within the A320 (Figure 4.10) shows the general location of the air-conditioning bays

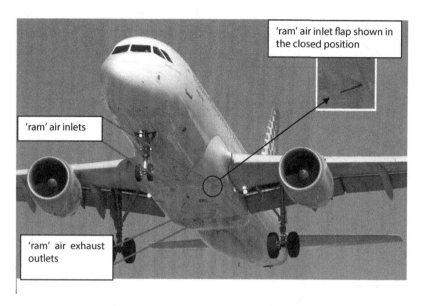

FIGURE 4.7

Photograph of the Airbus A320 taking off, with the indicated pack ram air inlets and outlets, including an enlarged inset of the ram inlet in the closed position.

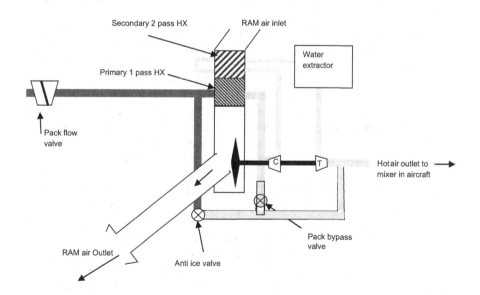

FIGURE 4.8

Schematic of the Airbus air-conditioning bootstrap pack operation when the aircraft is in-flight, providing heating to the cabin.

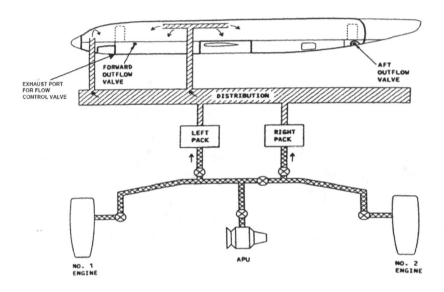

FIGURE 4.9
The A320 air-conditioning overhead panel, with the pack flow selector highlighted. (AAIASB.)

(one for each pack) and the distribution of the conditioned air throughout the aircraft.

The aircraft bootstrap air-conditioning system has a common vulnerability, namely that of the pack overheating in service. If the flight crew select a low desired cabin temperature under 'normal' flow conditions using the zone controller, and the aircraft structure is very hot from the solar gain operating in a sub-tropical environment, the pack controller attempts to achieve the required cabin temperature by increasing the bleed airflow to the pack. When high bleed airflows are ported to the pack, the turbo machinery responds and provides a cooler flow. However, this is dependent on the ram air cooling flow being capable of force-convecting the bleed air heat to the PFHE. If the outside air temperature of the ram air is too high, the PFHE will be unable to cool the hot bleed air sufficiently, which causes the pack temperature to rise rapidly. The pack controller attempts to counter this effect by porting a higher bleed airflow to the turbo machinery (i.e. a negative feedback effect). If the ram cooling and PFHEs continue to fail to cope with the demand of the cool cabin air requirements and the heat transfer limitations, the pack can enter a thermal runaway scenario (i.e. the pack temperature exceeds about 260°C). To prevent the pack and turbo machinery from a thermal runaway situation, a number of critically placed thermocouples are situated within the pack, which gives a warning signal to the flight crew via the flight deck indication system. If a thermal runaway situation is allowed to progress, the

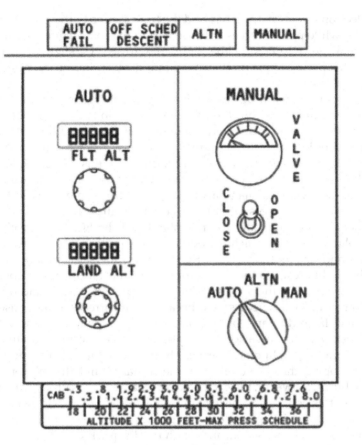

FIGURE 4.10
A schematic of the air conditioning and ventilation A320. (AAIASB.)

turbo machinery will rotate at such a rate that it will generate sufficient heat from friction to cause the turbo machinery to seize, or even worse, the pack components within the pack to combust. Flight crews are trained to isolate a pack that is indicating an overheat condition by closing the pack flow valve's switches on the overheat panel (Figure 4.9). If the aircraft is on the ground when an overheating occurrence takes place, the flight crew should return to the parking stand and seek an engineering action to rectify the fault.

The aircraft packs are usually operated with the pack flow selector (Figure 4.9) in the 'NORM' position, ensuring that the correct level of ventilation of the cabin (i.e. in a standard passenger configuration, namely a business and economy class) is achieved in-flight.

There are two air-conditioning packs fitted to the A320 aircraft, and a similar number on other commercial aircraft. Pack 1 typically supplies the flight deck, electronic and electrical bay and the front section of the passenger compartment, while pack 2 serves only the mid to rear passenger compartment. The normal procedure is for the flight crew to select a slightly lower temperature for pack 1 to ensure that the flight deck and electrical systems are adequately cooled.

In exceptional circumstances, such as in an emergency condition when the turbo machinery within the pack has seized in-flight, it is possible for one operational pack to be operated under 'hi' flow conditions to provide adequate ventilation to the passengers and crew prior to an unscheduled landing. However, continued flight at high altitude is not possible due to the lack of pressurisation, so flight crews would descend to a safe altitude (in accordance with the manufacturers' guidance) where a single ECS can provide sufficient services, prior to landing at an airport.

It is helpful for the reader to understand that the ECS is controlled by a Line Replaceable Unit, which is a dedicated computer within a casing in the electrical and avionics bay. This LRU interprets the selected input controls from the flight crew, providing the pressurisation levels and temperatures (within the scope of the systems) to the aircraft. An understanding of the operation of the ECS is necessary is because an incorrect setting or use of the system will have dire and catastrophic consequences for the aircraft occupants. There have been various examples in aviation accident events where the well-meaning flight crews turned off the ECS while the aircraft was still in-flight, hoping that it would slow the fire and limit the oxygen to that could be burnt. However, the grim reality is that the fire is unlikely to be suppressed by this action, and history indicates that if the ECS is switched-off, the passengers are incapacitated very quickly, breathing in the smoke and other products of combustion, and death results quickly.

When the plane continues to fly at high altitude, the operation of the ECS system must remain active, even in the most extreme failure scenarios that occur. Presently the sole control for this system is via the flight deck overhead panel (Figure 4.9), with the selection of the PACK valves being OFF or ON, and the flow rate switch being, LOW, NORMAL or HIGH.

The ECS provides a constant flow of pressurised air to the aircraft. The aircraft is constructed to be a sealed pressure vessel (e.g. a tube with ends), with controllable valves fitted, known as 'Out-Flow Valves', to release compressed air to the atmosphere. The Pressure Altitude (PA) that the occupants of the aircraft experience is based upon several factors. Assuming the aircraft is in-flight, the first factor is the physical altitude the aircraft is flying at. The next factor affecting the PA is the ECS mass flow (rate) from the packs into the pressurised fuselage structure. The final factor would be the position(s) of the Out-Flow Valves, which are controlled by a pressurisation controller/computer. Without the safe and effective operation of the Out-Flow Valves, the aircraft structure would simply inflate like a balloon, and burst when the forces of internal pressure exceed the manufacturers' tolerances. These tolerances are defined by the manufacturer, termed as Maximum Differential Pressure (max dP), and are the maximum difference between the outside local air pressure and the inside pressure. Clearly, an aircraft bursting open in-flight would result in a catastrophic event, thus additional levels of safety are incorporated into the pressurisation controller and the Out-Flow Valves.

Large, modern commercial aircraft have two or more Out-Flow Valves located at the front and rear of the fuselage, on the lower portions of the structure (i.e. below line the wings, Figure 4.9). The location of the valves in the lower part of the fuselage allows for the aircraft to ventilate the air directly from the under-floor cargo areas. The pressurisation controller ensures that the cabin differential pressure should never exceed the manufacturers' Max dP; thus, a safety margin is incorporated into the control programming. As a final layer of defence, many manufacturers include a mechanical override (similar to a spring-loaded valve) to allow excessive internal pressure to vent to the atmosphere- this releases the pressure prior to a catastrophic event, fully automatically.

These Out-Flow Valves can be manually controlled in very exceptional circumstances, such as forced landing on water (Ditching). This is because after the aircraft comes to a stop on the water, it would be important to prevent the Out-Flow Valves from allowing water to flood the aircraft, especially prior to an evacuation. Another possible emergency scenario that pilots train for is that of in-flight fire inside the aircraft. As this is such a significant event, flight crews would opt to land immediately. The smoke poses an immediate risk to life in an enclosed environment, as demonstrated in the British Airtours disaster in 1985 (engine fire at take-off, leading to cabin fire on the ground). In addition to fighting the cabin fire, the ECS would be operated at the maximum flow rates to provide fresh air to the passengers. While the aircraft was descending as quickly as possible, the Pilots could manually control the Out-Flow Valves to open them fully to eject as much smoke as possible, assuming the altitude is acceptable. Manual opening of the Valves can only be performed at altitudes less than 10,000ft, because if the cabin altitude exceeds this value, the passengers will also suffer the effects of the altitude (hypoxia), in addition to the possible exposure to gaseous products of combustion.

If the ECS is switched-off in-flight (at high altitude) or the Out-Flow Valves manually opened, this will result in the loss of cabin pressurisation with the subsequent deployment of the emergency drop-down oxygen masks. If the aircraft continues to fly at the same altitude, even wearing the emergency decompression masks will not be sufficient to survive. This is because these emergency oxygen masks only provide 'oxygen-enriched air' at local pressure, meaning that in the absence of a pressurised environment, the occupants will still succumb to hypoxia and incapacitation, as the partial pressure of oxygen is inadequate for human respiration. A good example of the equipment necessary for survival for low-pressure cabins are demonstrated with the US

FIGURE 4.11
U2 pilot walking to the aircraft wearing a sealed pressure suit attached to a portable ECS.
(Christopher Michel.)

Airforce operation of the 'Dragon Lady' U2 aircraft, as the pilots need to wear
fully pressurised flight suits (Figure 4.11 showing the pilot in the pressure suit
attached to a portable ECS), which were later used by NASA Astronauts.

4.4 Conclusions

In summary, the flight control systems for large passenger aircraft are powered movable surfaces usually relying on hydraulic assistance for precise
movement. With older aircraft, Boeing has prefered to incorporate a combination of metal cables attached between the pilots' yoke and the moving surface, coupled with hydraulic actuation to ease the forces necessary.

Airbus, in contrast to Boeing, has always incorporated a different philosophy utilising the 'fly-by-wire' technologies. Fly-by-wire implies that the various flight computers are calculating the magnitude of the flight surface position, based on inputs including the pilot's side stick input, among other data. Operationally, the vast majority of a flight is flown using the Autopilot function, regardless of manufacturer (e.g. Airbus, Boeing, or others).

The ECS is another fully automated system that pilots do not usually control manually, as the relevant computer controlling the system ensures normal operation. The ECS provides the life support necessary to keep the occupants of the aircraft alive. Again, the manual control of the ECS is only possible from inside the flight deck, but pilot intervention in exceptional conditions is possible, i.e. to increase or decrease the bleed airflows to the passenger cabin, and/or to control the levels of pressurisation via the Out-Flow Valves (manually).

The next chapter will explain why aircraft maintenance is important, and show how these activities are optimised by manufacturers and airlines, using newer systems to predict when individual components are likely to fail, providing financial savings for the operator.

5

Use of Live Aircraft Data in Aircraft Maintenance Management

5.1 Introduction

The use of live aircraft performance data has been demonstrated in recent years to be a very valuable tool to observe how individual components are performing and to estimate their remaining usefulness before their performance deteriorates to an unacceptable level. If a component is allowed to continue to operate below a given threshold, damage can occur to both the system and other nearby services. The resulting unplanned maintenance is very costly for airlines relying on their aircraft being available throughout the working day to carry revenue, passengers and freight.

This chapter includes an overview of the maintenance philosophies and provides an explanation of where these documents originate, and why reliability has commercial value. The sources of performance data are discussed. The use of wireless communications are explored, as are the advantages of the presence of big data predictive algorithms to estimate performance.

5.2 Aircraft Maintenance Management and Its Commercial Importance

Large commercial aircraft are complex devices that contain a very large number of complex components. In order for an aircraft to be dispatched into commercial revenue service, the aircraft must perform or operate to the approved tolerances that the original manufacturer has defined. For example, the engine manufacturer defines how many take-offs and landings an engine can make before it requires components to be changed.

Everything mechanical that moves, requires maintenance to remain serviceable. Applying this understanding to a large wide-body commercial aircraft with upwards of 10^6 individual components means, in the most basic terms, the minimum standards of performance must be reached before the aircraft can take-off to earn revenue.

Aircraft operators must fly these aircraft as frequently as possible to gain as much revenue during the working permittable flying hours of the aircraft each day. If night-time airport curfews occur (e.g. midnight), the industry practice is to conduct routine maintenance activities during the hours of darkness. This allows the aircraft to be ready for commercial service when the flying day commences (e.g. 5 am), making full use of the natural downtime of the aircraft.

5.2.1 Maintenance Planning by the Original Equipment Manufacturer (OEM)

When a new aircraft is designed, produced and flown for the first time, the National Aviation Authority of the country of manufacture becomes responsible for the certification of the new aircraft. In Europe, the European Aviation Safety Agency has legal powers derived from EU legislation, which is devolved to the NAA's. As part of this new certification process, the aircraft manufacturer must demonstrate to the Authority that a realistic minimum standard of engineering performance can be achieved. The aircraft OEM must also define the ongoing scheduled maintenance activities for a new aircraft with the certifying authority. These scheduled maintenance activities are usually conducted at predetermined time or use-based intervals. The defined maintenance activity is based on how the manufacturer foresees the aircraft being operated, and the type of inspection varies significantly. For example, a daily maintenance inspection may only have 20 or so checklist items, with weekly inspections being more detailed in terms of inspection items and complexity of test. An 'A check' inspection occurs at 4-month intervals, taking several days to complete, along with 'C check' hangar activities taking place every 18 months and being very complex. A typical 'C check' would see the engines removed and returned to the relevant manufacturer for strip and overhaul; the aircraft seats removed; floors removed; all the systems checked/tested, implementing any maintenance upgrades or modifications necessary; and the rectification of any other defects so that the aircraft is free from any Tech Log defects. A typical 'C check' duration would be 4 to 6 weeks of high-intensity maintenance in a hanger (working continuous shifts).

The maintenance schedule forms part of the aircraft's certification; the operator relies heavily on the OEM and documentation to define all the maintenance and the ongoing upgrades. The OEM business has many similarities to that of the automotive industry, with financial drivers being the principle factor. An aircraft OEM will communicate with all aircraft operators to offer maintenance, modifications and upgrades. The airlines have the task of categorising these communications from the OEMs, to prioritise modifications that will enhance safety and implement them quickly; this activity is likely to include the Airworthiness Directives of the relevant Aviation Authority to address known safety issues specific to a type of aircraft. The next type of modification will be cost saving, to allow an operator to run the aircraft more efficiently, and to make financial savings. The lowest level of modification will include all other possibilities to change the state of the aircraft, for example, the inclusion of onboard Wi-Fi data communication for passengers.

The airlines engineering office team must review all these communications from all the OEMs in a timely fashion, and quickly decide what is a *must have*, *nice to have* or a *don't need*. The OEM, being the manufacturer, will be more profitable if all the suggested modifications are implemented, whereas the airlines will bear the financial burden of both the activities and the loss of revenue availability of the aircraft.

5.2.2 Unscheduled Maintenance

If the aircraft develops a technical fault during the day, or if a fault is identified during the night time maintenance activity, these defects are entered in the aircraft Technical Log (Tech Log), a mandated record of all items that have arisen during the day or flight sector, and every registered aircraft must carry a copy of this document. Previously these Tech logs were 4 or 5 sheets of coloured carbon paper that were torn out before the flight with a white top sheet that remained in the book. However, in recent years, with the advent of cheap handheld computing and applications, the data is directly added by the person reporting the fault onto an electronic handheld device (e.g. an iPad or similar, containing specific applications), removing the need for large numbers of administration staff in the Engineering offices who previously would type out all the paper-based data into a computing maintenance database.

When faults arise, the details of the fault are recorded by the pilot or engineer into the electronic Tech Log. The airline that operates the aircraft must address the fault before the next flight in order to remain legally compliant with legislation. All maintenance activities must be performed by approved maintenance repair organisations (e.g. FAR 145/Part 145 approved companies) employing licensed maintenance engineers.

5.2.3 Minimum Standards of Equipment of Systems – Master Minimum Equipment List

When an aircraft is certified by a National Aviation authority, one important OEM document known as the Master Minimum Equipment List (MMEL) is produced and approved, and this document specifies what systems can be unserviceable, i.e. the plane could continue to fly safely with some defined defects. Some potential defects are quite trivial, such as a tail logo light filament that illuminates the painted artwork on the vertical stabiliser, since the maximum timeframe to defer and later perform the repair could be tens of days from the time of entry. However, other defects may be sufficiently serious and affect safety so that only one or two sectors can be flown.

If a defect occurs during the flying daytime, when the aircraft lands an engineer must evaluate the problem (for safety and legal compliance) and perform a recorded maintenance action in the aircraft technical log before the next flight. If the defect can be rectified quickly and the engineer has the correct spare part(s) available, the favoured option is to do the repair on the 'ramp' next to the terminal so as not to cause delays to the flying schedule. However, often repairs and testing take time and resources and a delay to the remaining days' flight schedule would have a significant impact on the operator's revenue, so a deferred maintenance action may be chosen by the engineer.

All defects are categorised by the ATA 100 code, an American 00 to 79 code classification of all systems for fixed-wing large aircraft. The engineer would check the restrictions written in the Minimum Equipment List (MEL), an airline-specific document that is more restrictive than the OEM's MMEL. Providing the defective item can be deferred, the necessary administrative actions are undertaken and the 'open' failure is 'closed' by the engineer, and an Acceptable Deferred Defect (ADD) is raised. The number of ADDs is carefully monitored by the Engineering management and closely scrutinised by the National Aviation Authority, as the emphasis is to use this activity only where appropriate. Typically, if the MEL permits, AADs can last up to seven consecutive days, and if the defect cannot be fixed, the item can be 'closed' and 'reopened' twice more, giving a total timeframe of up to 21 days from the initial entry.

If the aircraft defect can be deferred to a 'C check' the operator will defer the rectification until the scheduled aircraft planned hangar maintenance check (because the aircraft will be in the hanger for up to 6 weeks), and this is known as a Base Deferred Defect (BDD).

If the defective system has an effect on the safe operation of the aircraft or is not permitted to be a deferrable item within the MEL (i.e. the permissible defects and limits), then the aircraft is grounded until such time that the defect is fixed, and the engineer approves the repairs to a standard *In Accordance With the Aircraft Maintenance Manual* (*IAW AMM*). The industry now categorises the grounding of the defective aircraft as Aircraft On Ground (AOG), and this code is the highest level of maintenance priority.

5.2.4 Component Reliability and Maintenance Strategy

Unscheduled defects pose a real problem to the efficient operation of the aircraft because it is very challenging to 'predict' when the next component might fail.

Reliability is often considered as the measure of serviceable performance with time. If a component performs in such a way that is outside of the original designed specification performance, it is considered to have failed. Reliability as an engineering field is a mature subject, with roots stretching back to the Second World War. During the Second World War, mechanical systems were expected to have a finite lifespan, due to assumed damage that would naturally result from conflict. The logic was that a system would require less regular maintenance because the asset, be it an aircraft, land vehicle, etc., would have a given probability that it damaged and potentially lost or unrepairable. Consequently, emphasis was given to manufacturing items in very large volumes rather than for long-life performance. When maintenance was performed, all components on a vehicle removed down to a single-component level and inspected. If necessary, additional maintenance and new components would be fitted. This type of philosophy, i.e. very labour-intensive maintenance, tear-downs, inspection and rebuilds continued up to the early 1970s. This is reflected in the duration of a licensed Aircraft Engineers maintenance course for a Boeing 747-100, which took some 3 months for the engine and airframe elements. By comparison, a maintenance course of similar scope for a B747-400 series is around 4 to 5 weeks. The difference is explained by the change in the maintenance philosophy between the very early 1970s and the present day: the regular maintenance activity of the present time has been optimised for the use and the operation of the system. The need to 'pull everything apart' has been superseded with a more mathematical approach, based on performance data for large numbers of components.

5.2.5 Bathtub Curve for Reliability and Mathematical Predictions

The concept of reliability considers that individual components all have differing levels of wear and serviceability. A mathematical evaluation of a large number of components in service and their time to failure has been established for many years, known as the 'bathtub curve' as illustrated in Figure 5.1, where the Y-axis represents the failure rate and the X-axis the time elapsed to failure.

Figure 5.1 shows three distinct regions of the reliability *Bathtub* shape, namely the curve to the far left, the curve to the far right and lastly the flat line between the two curves. The left zone (Figure 5.1) is known as the *infant mortality* region and the decreasing curve value represents the manufacturing defects that might result in an earlier than expected failure. Conversely, the right-hand side (Figure 5.1) is known as the *wear out* region and is attributed to the increasing failure rate due to the item wearing out more rapidly, and the curve illustrates the exponential increase. The flat line region (Figure 5.1) illustrates the random failure rate of the majority of components, as the parts are engineered to perform for a minimum time duration and reach the right side of the curve.

Mathematical evaluation of the time to component reliability can be performed, such as the Mean Time To Failure (MTTF) or Mean Time Between

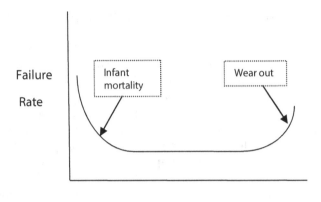

Time

FIGURE 5.1
Reliability *Bathtub* curve of component reliability.

Overhaul (MTBO). The mathematical models are valuable because they are based on a large data set for identical components. If the data set is not sufficiently large and several *infant mortality* events are experienced, the data arising from the *Bathtub* curve would show the mathematical MTTF as being much smaller than the *real-world* large data set. The objective of the reliability calculation is to predict when the average or MTTF will occur, and ultimately to remove the component before possible catastrophic failure, overhauling it to *like new* performances.

The use of these mathematical tools to evaluate performance permitted the exchange of components to be factored into the overall maintenance strategy, which is dictated by the OEM. If an item cannot be reasonably repaired, then it is classified as a consumable part, and after use is disposed of by the appropriate means. However, many of the components are complex and expensive, so, the justification to repair them at an approved company is both justified and necessary. These components are known as rotables – i.e. they can be repaired by a third-party approved overhaul organisation, re-certified as *good as new* and re-fitted onto live aircraft. Clearly if the rotable parts are operated until catastrophic failure, the repair and recertification is not going to be physically or economically possible, especially if the components have structurally failed. Accordingly, reliability tools are utilised to predict the optimal usage time for each component.

An aircraft OEM will optimise the maintenance programme for their aircraft, including the use of reliability software to predict the given components' failure rates, as the overarching goal is to keep the aircraft flying as long as possible. The aircraft OEM also minimises the maintenance to acceptable levels by the certifying authority such as the Federal Aviation Administration (with the goal to minimise *unscheduled* maintenance) and to schedule all component changes, where possible, to planned scheduled maintenance, namely the 'base C' checks in the aircraft hangers.

This complex situation of planning, predicting and monitoring the status of both the aircraft and the tens of thousands of individual rotables and components is only possible with the use of computing.

5.3 Technical Components Combined with Data Logging

Historically, the early forms of data logging came from human-based observations that would be recorded in paper-based systems, such as the technical log. For example, during each of the engine starts, the pilot would carefully watch the indicators in the flight deck, and record the temperatures during the moment at which the engine accelerates when a second set of fuel injectors opens, in addition to the post-start ground idle Exhaust Gas Temperature (EGT) values. These values would enable an office-based technical service engineer to compare the engine performance, with additional data (such as engine use since overhaul, oil lubrication consumption, etc.) to estimate the optimal time that the engine should remain in service. An excessively worn engine consumes much higher quantities of fuel and/or oil, produces less thrust and operates at hotter temperatures, while the internal components wear-out much faster. Based on these data collected, the staff in the offices need to carefully balance when it is most effective and commercially advantageous to change either the engine or the relevant components. However, it is worth noting that the paper-based data recording and evaluation system would always be several days late due to the paper processing and normal working office hours. Moreover, if any long weekends or public holidays occured, the closure of the usual office working hours added up to seven days in the delay between the paper recording of the data and the electronic interpretation of the performance.

With the development of the DFDR technologies, it became possible to download this operational data via non-protected recorders known as acquisition units. Early technologies used a device known as a Quick Access Recorder (QAR) (Figure 5.2).

The QAR allowed for the data to be transferred to a removable cartridge, thus allowing the data collected at a given interval, and then the cartridge's data (Figure 5.3), to be uploaded to a database. The drivers for this change were the improvements in technologies (i.e. data storage), but the real driver was the ability to collect live flight data to better optimise the maintenance schedule; hence it was financially driven. If an operator can delay an engine change by a few days to fly the aircraft in revenue service, this will result in a financial gain for the operator. If the maintenance was extended too far, then

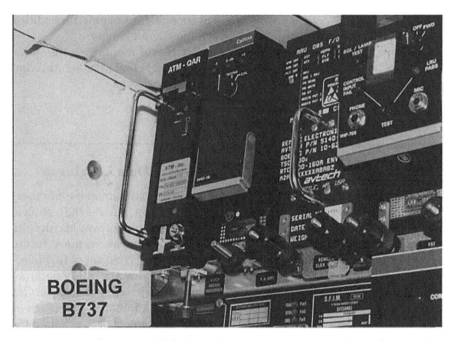

FIGURE 5.2
Boeing B737 (classic) Quick access recorder module. (ATM PP Ltd.)

the engine or component may fail, resulting in unscheduled landings and significant extra expense. Thus the duty of reliability engineering and the predictive optimisation of maintenance was shown to be worthwhile.

The limitations of earlier recording devices included the storage capacity of the aircraft-based device. Early units had limited resolution (the ability to record a given data set per unit time) and may had to be downloaded once a day. However, technology and recording capabilities have improved exponentially since the 1970s: operators in the late 1990s were able to download a weeks' worth of on-wing performance data each time, making the retrieval of the data more effective in terms of man hours. In 1999, it was possible to download the flight performance data with a laptop (or desktop) computer module using cables that were connected to the acquisition unit in around 25 minutes (Figure 5.4).

While the recording of data quantity had improved up until around 2001, the task still required a licensed aircraft engineer to book the laptop and cable from the stores, go to the aircraft and download the data weekly.

On 11 September 2001, there was a significant act of terrorism in which multiple commercial aircraft were flown into buildings across the USA. The importance of this event was later reflected in the financing of plane engines. Before 9/11, the majority of carriers leased their aircraft, and likewise leased

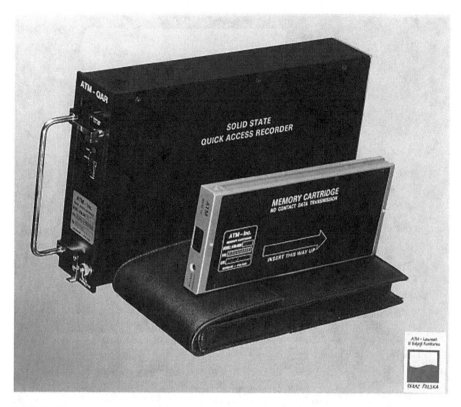

FIGURE 5.3
The Quick Access Recorder module and memory cartridge. (ATM PP Ltd.)

the engines fitted to these aircraft. The drawback of leasing an engine is that the leaseholder (i.e. the airline) is liable for the financial costs of the engine, regardless of whether it is working or not. If the engine was being overhauled at a repair station, the lease payments were still liable, and the payments continued until the lessor agreed to accept the asset back from the creditor, and only at this time did the payments stop. The financial liability of airlines post 9/11 was immense, and many airlines changed their engine leasing arrangements with manufacturers to a different financial purchasing arrangement, known as 'power by the hour'. Effectively, the engine remains the property of the manufacturer, and the carrier pays an agreed hourly rate for the operational engine; thus reliability concerns held by carriers were shifted to the engine manufacturer. Engine manufacturers circa 2001 quickly established modern communication equipment, and in order to better maintain their engines they required a new means of technology to monitor engine performances. The engine, after all, is the asset of the manufacturer: how many

FIGURE 5.4
A desktop computer and cable capable of downloading the data from the DFDR. (ATM PP Ltd.)

licensed aircraft engineers, working for the likes of GE, Pratt and Whitney or Rolls Royce are available at a commercial airport?

The simple solution chosen by many engine OEMs was to use the data storage that was part of the Full Authority Digital Engine Control (FADEC), a computer device to control the gas turbine engine. The FADEC records all the operational parameters of the engine, in addition to other performances (altitude, airspeed, temperature, power requested, etc.). The data were obtained by the manufacturer by fitting a communications module to the FADEC. A GSM phone attached to the FADEC would dial the manufacturer's server when the aircraft was on the ground and the park brake was applied, thus allowing the transmission of the data via GSM networks. Many of the major OEMs offering *Power by the Hour* mandated the fitting of this technology as part of their terms of agreement. The introduction of this communication module allowed the manufacturers to access all the flight performance, trends. This has proven to be a very valuable tool in ensuring that the asset is used correctly, and for maintenance to be optimised.

5.4 Live Streamed Data and Radio Communication Technologies

5.4.1 ACARS

In 1978, a new ground to air communication system known as Aircraft Communications Addressing and Reporting System (ACARS) was developed by the Aeronautical Radio Incorporated company as a tool to send data from the ground to the pilots without the use of voice communications. Initially, ACARS services used VHF or HF radio communication, producing paper printouts (Figure 5.5) in the flight deck using thermal paper that looked very similar to till receipts or fax machine printouts.

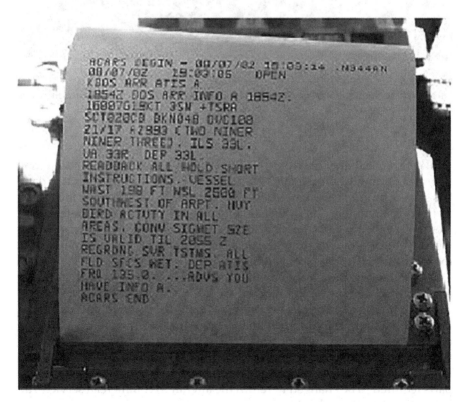

FIGURE 5.5
ACARS thermal paper print out.

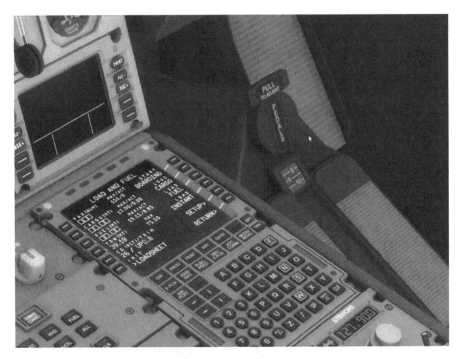

FIGURE 5.6
Photograph of Airbus Multi-Function Control and Display Unit – the interface including ACARS interface.

Likewise, the flight crew could send text messages back to the base station (Figure 5.6), and these systems were able to interface with the aircraft's data-bus system or use predesignated functions to request the weather forecast or company communications. As radio and satellite communication technologies improved, it became possible to send and receive ACARS messages anywhere in the world, including the polar regions, which had previously been too far from VHF radio base stations.

ACARS became more useful to the manufacturers and operators as newer aircraft (B747-400 onwards) allowed the ACARS system to fully interface with monitoring equipment. For example, if an Airbus A320 aircraft system developed a fault in-flight, the databus would record the failure in the aircraft recording system and the ACARS would automatically send a message to the server in Toulouse. Toulouse server would then relay the details of the failure instruction to the operator, and would include details of the necessary parts required to be changed to return the aircraft to commercial service – all before the aircraft had landed.

Additionally, ground-based engineers could use a PC terminal connected to the internet with proprietary software to communicate, via the ground-based server, with any aircraft in the world and to interrogate the relevant system. The author has seen first-hand this technology in use in 2004, when he was engaged in research regarding the failure of air-conditioning systems. With the support of a local Airbus technical representative's kind assistance, the author was able to view the in-flight performances of various Airbus aircraft that had recently departed London Heathrow airport for the USA. None of the data requested from the aircraft required any human interaction from the flight crews of these aircraft, rather the request was made to view the air-conditioning system performance, and some three to four minutes later the data were displayed on the terminal screen. The purpose of such a ground-based interrogation was explained by the technical representative 'as being able to troubleshoot those difficult technical issues'. Previously, for some maintenance problems, the only method of fully rectifying the failure was to remove the aircraft from the flight programme, and for a licensed aircraft engineer to take the aircraft, with flight crew, for a test or maintenance flight. The financial implications for this activity were significant: thus the integration of the ACARS capabilities into the aircraft's technical recording system was financially driven.

The weakness of the ACARS system is the subscription-based nature of the service to the data carrier, as well as the quantity of data packets that can be transmitted via VHF transmission. The principle of a robust predictive maintenance system is the availability of data stored in Very Large Data Base's (VLDBs) that can be analysed using mathematical trends and computer calculations. Throttling the data transmission due to the constrictions of the VHF characteristics effectively limits the use of ACARS as a tool to transmit the full data collated from the aircraft to the ground servers that form the VLDB. Additionally, the costs of transmission for a full onboard system, broadcasting all data parameters, would be higher than what the aircraft carrier would consider to be acceptable. An example of this lean approach to data management and subscriptions was highlighted by Malaysian Airlines after the MH370 disappearance. The Malaysian Airlines B777 aircraft were fitted with both VHF and Satellite ACARS communication equipment, yet the subscription, which was around $10 USD per day per aircraft, was deemed too high, so the subscription for this service was not undertaken. In contrast, Air France used this subscription service. When the AF447 aircraft crashed off the coast the of Brazil on 1 June 2009, the transmission of flight information via the satellite systems was instrumental in locating the ditched aircraft on the seabed. While it was not possible to immediately identify the precise location of the aircraft on the seabed, a general region was provided from aircraft to satellite data. This allowed for a smaller search area to be established, and on 3 April 2011, the debris of the aircraft was located at a depth of 4,000 m.

5.4.2 British Airways Engineering Maintenance Management

British Airways understood the need for VLDBs to perform robust data management predictions that enhance maintenance operations. The first significant investment was around the year 2000 with the Enterprise-Wide Scheme (EWS) IT solution, contracted to EDS: some 120 separate IT systems within BA were integrated into a single system that could be interrogated by standard desktop computers (rather than bespoke terminals). In the two years that followed the introduction of EWS, the IT system matured, and single solution system became the backbone of BA international operations. The second, more ambitious activity coincided with the completion of Terminal 5 at London Heathrow airport, as all BA aircraft operate from this hub. The plan for data management centred around fitting all the parking stands with *brouters* (i.e. wireless communication ports). Additionally, BA retrofitted their entire aircraft fleet with wireless communication cards to automatically transmit operational data to their own computer servers. The principle of the British Airways system is to use the onboard data storage previously mentioned. When the BA aircraft parks at the Terminal 5 stand and applies the parking brake, the wireless communication device activates, connects to the *brouter* and downloads the stored data in its entirety to the BA server. Another difference with the BA maintenance management solution is the ownership of the data. The data are recorded and acquired by the carrier, namely British Airways. In other commercial solutions, the data are acquired by the ACARS subscription service, transmitted to the OEM servers (i.e. Airbus, Toulouse) where the operator may view their own airline's performance data on the Airbus server.

The UK government recognised this significant investment and cited BA's initial set-up costs at £150 million, to make better decisions on out of service dates and economic repairs. Furthermore, the system is reported to have made British Airways a cost saving of around 30%, as detailed in the UK government's report (Transforming logistic support for fast jets, National Audit office 17 July 2007).

5.5 Data Mining of Very Large Data Bases and Commercial Solutions to Predictive Maintenance

5.5.1 Rolls Royce Commercial Engines

Commercial engines are very complex units: their tolerances for performing efficiently and effectively are very small. One example is the combusters, located in the engine where Jet A1 fuel is injected in a vaporised sate. Once ignited and fully burned, this is naturally the hottest location in the gas turbine. The problem, however, with such high-temperature locations is the material science. The very high temperatures, along with rapid changes in temperature and gas flows, result in the combustor components being subjected to extreme wear, resulting in failure. Over the last 50 years, new applications for materials and coatings have led to improved on-wing performances, but the limitations of this science are very apparent, with the need to regularly inspect such hot areas, recording all visible defects. The engine manufacturers design the engine to perform with a given mass to pre-defined tolerances and service intervals. Therefore, if a new engine for an aircraft is too heavy (exceeding the airframe OEMs initial specification) or does not produce satisfactory on-wing performance (because it requires component changes), there will be a detrimental financial effect on the engine OEM..

Rolls Royce estimates that the total operational cost for an aircraft turbine engine is approximately 4% of total costs, represented in Figure 5.7. The high cost of engine operation has always been a great concern for the aviation business. Rolls Royce understood the need to follow engine performance trends for many years, and in the late 1980s onwards Rolls Royce worked very closely with Cathay Pacific Airways and American Airlines to better improve performance and reduce operating costs. In the UK Derby site, a new mathematical approach was led by Dr Michael J Provost (PhD December

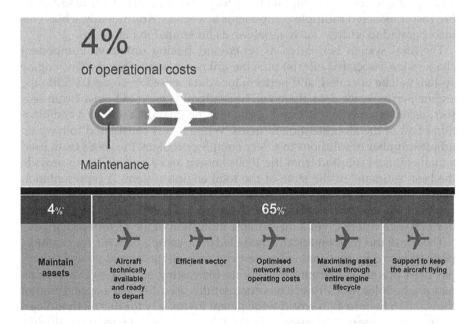

FIGURE 5.7
Rolls Royce estimation of the engine maintenance costs versus total airlines operating costs. (r2pi.)

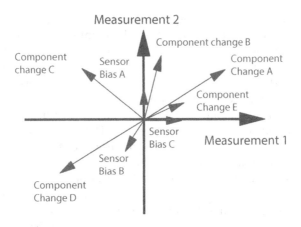

FIGURE 5.8
Rolls Royce correlation between component changes and sensor bias. (Rolls Royce.)

1994, Cranfield University), to utilise a new mathematical approach to predict engine performance. Provost recognised the difficulties with the infinite possibilities of variables with the engines, especially if components are removed and replaced (thus introducing new elements of change). However, instead of considering component performance as a linear representation, Provost considered multiple components as vectors. Additionally, a bias was incorporated to address such variables, as illustrated in Figure 5.8.

The total system is defined as an engine, having unknown component changes (as associated effects) plus the unknown sensor biases. This engine system will be operated, and performance data sensed/recorded. Within the engine performance data, there will be randomly generated noise (from sensors, recording media, etc.), and ultimately the final data will be a combination of all of the aforementioned items. While this is considered to have an infinite number of solutions to a very complex problem, Provost's team used a mathematical solution from the 1960s known as a *Kalman filter*, to provide the best 'estimate' of the state of the total engine system. A representation of this predictive system is illustrated in Figure 5.9. Blocks 'B' and 'C' are compared to the values of the ideal performance, and this variance between performance values gives a measurable indication of the recorded changes.

The use of this mathematically modelled data, compared with, for example, performance data captured with a test engine in flight-laboratory conditions, has been pivotal in permitting accurate forecasting and predictive maintenance schedules. Rolls Royce has successfully used Provost's application of the Kalman filter to gain world-wide patent exclusivity for this methodology, because the value of estimating equipment longevity is so financially valuable. Initially, Rolls Royce partnered with a Canadian organisation in a joint venture to form DS & S, which was the predictive maintenance holding company.

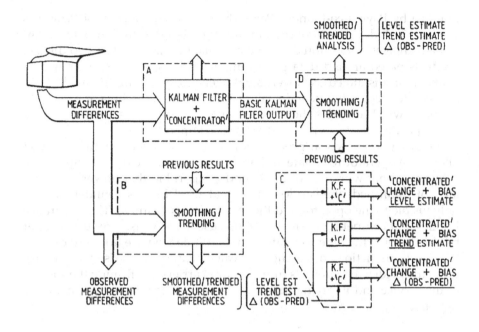

FIGURE 5.9
Flow diagram illustrating the application of the method to estimate levels and trends in Rolls Royce predictive maintenance tool. (Rolls Royce.)

This recognition of the value of reliability was further demonstrated in 2006, when Rolls Royce purchased all the shares of the joint venture and became the sole owner of DS & S.

Rolls Royce has recognised the importance of predictive maintenance capabilities, as although it values initial equipment sales, the income generation from service support (parts, maintenance management) has much greater financial longevity, which can be predicted by the industry up to 50 years from point of manufacture. This ultra-long-term ambition is only made possible by combining the use of embedded sensors within the engine and the ability of to retrieve live 'on-wing' data communications (via ACARS with satellite communication), so that real performance can be compared with the VLDB to optimise and achieve maximum performance from the on-wing asset.

5.5.2 Airbus, Palantir and Skywise

As previously mentioned, it is noted that all the major OEMs have commercial offerings for predictive maintenance. Airbus has partnered with a US-based software analytics organisation, Palantir, to capture and analyse aircraft performance data to form a VLDB that can be analysed. Partly founded by Paypal co-founder Peter Thiel, the early function of Palantir's products included counter-terrorism strategies designed to disrupt the financial advantages of electronic commerce. Palantir currently has two distinct products based around data processing algorithms. The first is known as Gotham, and is aimed at Government agencies, including the intelligence services, to map, network and analyse VLDBs. The second major product is Foundry, which includes their proprietary algorithm. This is the analytical technique that the 'Skywise' product is based upon.

The Skywise analytical product is marketed as a confidential tool that allows the operator to collate and analyse performance data (Figure 5.10). While the inner workings of the Foundry algorithm that powers Skywise are not published, the service states that the VLDB is confidential and only visible to the operator and the manufacturer (Airbus). This assurance implies that competing airlines that also use the Skywise product will not be able to view the other operator's data. While Skywise does not explain how such predictions or savings are calculated, the Skywise website has numerous video testimonials from prestige carriers, confirming the success of the service and implying the financial savings that have been made by the carrier.

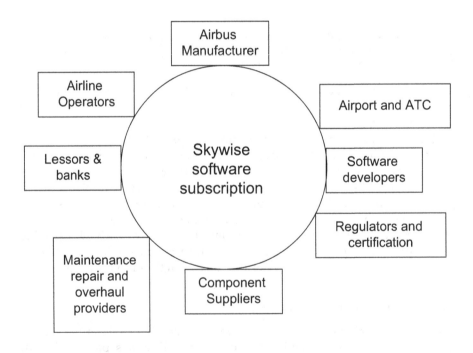

FIGURE 5.10
Airbus and Palantir's Skywise data analysis indicating all the actors.

Skywise relies on vast quantities of data being captured, and Palantir suggests that this collection of data can be measured in pentabytes. The system records operational data values from up to 20,000 different sensors per individual aircraft, with a recording frequency of between 20 and 100 values per second. Each flight is expected to generate approximately 1,000,000 data values per sector, which for a European operator is approximately 1 TB per sector. Operationally, because of the vast data sizes acquired and stored within each aircraft, the operator still requires an engineer to visit each individual aircraft with a laptop to download the stored data. Once in the office, the recovered data can be uploaded to the Skywise server for analysis.

Skywise claims (at the time of publishing) that over 100 airlines are contracted to the service, with 9,000 + aircraft being monitored and a VLDB in excess of 10+ PB.

In addition to the predictive maintenance function Skywise performs, the Skywise service is also able to compare previous operational data to allow for accurate forecasting of future operations, including mass and balance, fuel planning, and revenue management. For partner component OEMs, the service also allows for an accurate Root Cause Analysis of in-service components that would previously only be seen during the removal and return for overhaul to the OEM. Now the respective OEM can review how their components are performing/operating on-wing.

While Skywise states that the operator can choose what data to share with Airbus, and what to withhold, it is also helpful to note that Palantir participated in the British Parliament Inquiry in 2018, as Palantir had 'operational connections' including meetings with Cambridge Analytica (CA) with confirmation in the inquiry that employees of Palantir and CA had been sharing offices with other third parties. This admission by CA and others to the parliamentary inquiry highlights how extremely large data sets that are collected can be shared, sold or accessed by third parties, without the knowledge or approval of the end user.

5.6 Conclusions

The concepts of aircraft maintenance have been explained, including the differences between scheduled and unscheduled maintenance, in addition to ongoing maintenance checks. The use of aircraft performance data is a very important and valuable tool for the Airframe OEMs, component OEMs and lastly the airlines themselves. The origins of data collection lay in logging entries into paper-based engineering Technical Logs that were completed before and after each flight. However, paper-based systems required the data to be transposed into a very large database, and these practical delays resulted in faults in monitored items going unnoticed and working beyond their intended operation. Later, a new source of flight performance data was brought in with the DFDR, and new devices (Quick Access Recorders and Cartridges) were introduced to transfer the information from the aircraft to the computer database.

With modern aircraft in the 1990s, ground stations could request the airborne communication system (ACARS) to transmit a small number of aircraft system performances in-flight. This system performance was fully autonomous, and did not require assistance from flight crews. Another development was replacing QARs with the introduction of portable laptops with cables and modules, with the ground engineer regularly downloading the information onto the computer's hard drive.

After the events of 9/11, it became more practical for airlines to consider leasing operational time from engine manufacturers in a 'power by the hour' agreement. However, the engine manufacturers needed to remotely monitor the ongoing performances of their assets, so, data communication modules were added to the engines' FADEC computers. Rolls Royce developed their own predictive reliability software, and patented it – as did the other engine manufacturers.

Lastly, the collection of big data by the Airbus Group in conjunction with Palantir has demonstrated that significant levels of financial savings can be made. Some of this data is transmitted by ACARS. However, the vast majority of the system performances are still being transferred once a week using a laptop, with the data uploaded to a server at Airbus's and Palantir's facilities.

The next chapter will explain the background of human error, or 'Human Factors'. It will explore important events that are now used as case studies to demonstrate the need to improve safety management, and why this is a mandatory subject for airline staff.

6

Human Factors and Safety Management Systems

6.1 Relevance of Human Performance and Safety Management Systems

The subject of human performance and error, and its subsequent evaluation post-aircraft accidents, will be discussed here in detail. Included is a background in factors that contribute towards mistakes, and how employees in the aviation sector have made errors. The three basic Human Factors (HF) models are presented, including their usage and limitations, and why Human Performance became an important discipline. This includes the effects on human physiology that have contributed to accidents or significant events.

In addition to understanding the importance of HF and Human Performance, it is important to include the principles of Safety Management Systems (SMS) that are mandated in the aviation sector, including their limitations. While the SMS is not a perfect system to prevent accidents, it is a useful tool or process in mitigating events.

The principal purpose of having an awareness of HF is to understand the industries' wants and needs in regards to learning from their past experiences. As history has demonstrated, aircraft have crashed and lives have been lost. The desire to prevent a similarly tragic outcome has always been the driving force for improvement. Within airline companies, the emphasis is equally placed on their own bespoke SMS, which supports staff in performing the job correctly the first time, every time. Naturally, this in turn provides a significant financial advantage to the operator.

References to accident events are based on the open-source accident reports that are freely available to review.

6.2 British European Airways Accident – A Turning Point

On 18 June 1972, an aircraft crashed close to Heathrow Airport, killing all occupants. This event would go on to change the way accidents would be evaluated. The fateful flight was a British European Airways (BEA) Hawker Siddeley Trident aircraft, bound for Brussels. It crashed very soon after takeoff, claiming the lives of all 118 passengers and crew (see Figure 6.1). The aircraft crashed at 1610 hrs (British Summer Time) some 133 seconds after the take-off roll began, crashing into the suburbs of Staines town, now known as Staines-upon-Thames, to the west of London, UK. All occupants succumbed to massive trauma upon an impact caused by a deep stall. The aircraft, having gone into superstall, reached significant vertical speed and 'pancaked' on impact.

This event has become a standard training scenario, known in the Aviation Industry as the Staines crash. An analysis of the crash, and the events leading up to it, explains how tragedies such as this one can be prevented. While this section is not a full précis of the (Air) Accident Investigation Board (AIB in 1970s, AAIB in later years) report, some of the causal factors are explained and considered in a modern context, including the emergence of an error model.

At the time of the event, there was a significant ongoing trade union dispute between the British Airline Pilot Association (BALPA) and BEA, with

FIGURE 6.1
Hawker Siddeley Trident aircraft in BEA livery at London Heathrow Airport. (Malcom Taylor.)

the threat of strike action. It is helpful to consider the career backgrounds of the typical pilots that flew commercial aircraft at the time. The older, more experienced pilots had typically served in the military for considerable periods of time before transitioning to the commercial market. Many of these ex-military pilots had seen combat action in the various events between the 1940's and 1970s, and thus had a disciplined military thinking and attitude based on their personal military service experiences.

The younger generation of pilots entered the aviation sector having gained their private pilot's license flying in small 2 to 4 seat piston-engine aircraft. They transitioned to twin piston engine aircraft with 4 to 6 seats, built their flight experience with small aircraft before being recruited by airlines and gaining commercial experience. Naturally, the attitudes, beliefs and methodologies of the younger flight crew differed to the older, more military-focused ex-forces pilots. At the time of the Staines crash, the younger pilots were more aware of the possibility of potential strike action that was being threatened by BALPA in a dispute over pay and conditions because renumeration is the principal reason for many to go to work. Coming from a military background, the older pilots did not share this view, with strike action being something they were uncomfortable with. In short, there was a tense cultural division at the time between and young and old pilots.

This BEA 548 on the day in question comprised three flight crew member operating the aircraft: Captain Stanley Key, a veteran of the Second World War with strong, traditional views (left-hand seat P1); Second officer Jeremy Keighley (right-hand seat P2); and Senior First Officer Simon Ticehurst taking the role of flight engineer (Engineers seat P3). Additionally, the fourth jump seat was occupied by Captain John Collins. The relevance of the flight crew complement became apparent after the accident. From a skills perspective, the captain was the most experienced on the flight deck, but was feared by his colleagues. The junior members of the flight crew were keenly aware of captain Key's view of the impending industrial action. On the day of the crash, Key's was observed having a tremendously violent argument with another colleague regarding the impending strike action. Captain Key made his anti-strike views known, to the extent that this violent argument was witnessed by numerous other colleagues at Heathrow Airport. The working atmosphere during the briefing would have been tense, and was the first potential area for a problem to occur, as cited in the accident report.

The Trident aircraft is a T-tail jet aircraft that has traditional flight characteristics. The problems with T-tail aircraft are to do with the stall characteristics: namely if the aircraft enters a deep stall, the flight crew will be unable to recover the situation. This is because the elevator surface has become ineffective due to the stalling turbulent air passing over the tail section; thus the usual nose-down attitude required to recover is ineffective. To prevent this type of stalling event transpiring, the manufacturers added a stall warning system, comprising two angle of attack sensors on each side of the nose of the flight deck. In addition to the stall warning system, a stick push device is attached to the elevator control circuit, and this would typically activate after 10 seconds of continuous stall warning. An underlying problem with the Trident aircraft was that of false stall warnings during its previous operational service. A number of operational BEA crews experienced this spurious stall warning activation. Consequently, it was considered acceptable to cancel the warning and to continue flying, treating the warning system with some level of scepticism.

114 seconds after take-off, the stall warning system activated. At the time of the accident, flight deck data recorders were fitted to this model of particular aircraft, but the cockpit voice recorder was not a technology that was fitted at the time of the accident, as detailed in the AIB report. It is believed that as the stall warning sounded shortly after take-off, the flight crew cancelled this indication (the audible alarm, etc.) and attempted to fly the aircraft as per usual. Multiple stall warnings, stick shake indications and multiple stick push activations took place, yet the final outcome was the deep stall and impending crash.

Unfortunately, the aircraft was flying slower than it should have been. One contributing factor was the noise abatement procedures for aircraft flying close to the airport. These noise abatement procedures require the pilots to throttle back the engines, using less engine thrust as the aircraft passes out of the airport perimeter over the boundary fence. The logic behind this noise abatement procedure is to mitigate noise pollution for residents living close to the airport. Unfortunately, the reduction in thrust from the three engines resulted in a slower acceleration of the aircraft to the higher safe cruising speeds. In short, the aircraft was flying slower than intended, some 60 + knots slower than the action called for. Additionally, the leading edge and trailing edge devices were individually controlled: a characteristic unique to the Trident aircraft at the time. The accident investigation report determined that the leading edge droop flaps where retracted too soon into the flight (i.e. the surface was moved at an incorrect airspeed, being too low), and this premature action resulted in the loss of lift and the aircraft then entering a stall type condition, which later developed into a deep stall.

Large commercial aircraft have two or more members of the flight deck crew to better manage the workload. The Pilot Flying (PF) the aircraft, meaning the person controlling the autopilot heading, speed, altimeter, etc., in this case was P1 Captain Key. Meanwhile, the Pilot Not Flying (PNF) was Second Officer Keighley as P2, who would be required to monitor the critical values of performance, such as airspeed, rate of climb, etc., and would call out to the colleagues the required performance values, so the PF can then perform an action. Additionally, the Flight Engineer, Ticehurst, would be monitoring the technical performance of the aircraft and engines, and would ensure that the system as a whole was performing correctly. He would also be in charge of trimming - making very minor adjustments to the aircraft to enhance performance.

To a lay person reading the Staines accident report, it would seem impossible that three highly trained professional pilots could make such poor decisions, regardless of their differing levels of experience. However, explaining this was the challenge that the AIB investigation team faced at the time. The investigation focused on all the typical aspects of accidents at that time, with additional focus on the aircraft systems. The crashed aircraft and engines were removed to Farnborough, UK, the headquarters of the AIB. Analysis of the physical evidence from the crash site indicated that the aircraft did not suffer a catastrophic mechanical failure immediately prior to the crash. Rather the culture, the attitudes and the hierarchy affected all the pilot's performances to the detriment of safety. Additionally, the post-mortem medical examination of all the occupants indicated that Captain Key had undiagnosed heart disease, recorded as atherosclerosis. Furthermore, the confrontation at the airport between Key and another pilot was recorded as being the most violent confrontation that other pilots had ever witnessed at work, thus putting Key and his heart condition under extreme stress. The coroner recorded the evidence from the pathologist showing Key would have suffered heart problems not more than 2 hours before the flight, and not less than one minute before the crash. One significant hypothesis that continues to this day is whether Captain Key was suffering a heart attack at the time of the take-off roll, and his deviation from the standard protocol (including Radio Transmission) was indicative of his extreme medical condition. Furthermore, the problematic cockpit culture indicated the levels of dislike between the junior ranking staff and Key, who was seen as an 'old-timer'. The crashed aircraft had evidence of graffiti written on the flight engineers' desk regarding Captain Key. Likewise, other BEA aircraft were found to contain similar graffiti in the flight decks, indicating the culture was not unique to the operational crew of BE 548. Again, the P2, P3 and possibly even the observer on the flight deck of BE 548 would know all too well that this aircraft was not accelerating to the required speed. The stall warnings should have been a very significant message to all occupants of impending danger, yet no radio transmissions were made. While the investigation and subsequent government enquiry attempted to explain why this terrible event transpired, in order to prevent accidents such as this event taking place a new approach was needed.

In the period leading up to the Staines crash, Dr. Elwyn Edwards was a psychology academic specialising in ergonomics at the University of Technology (now known as Loughborough University), Loughborough UK. Edwards was no life-long academic that exclusively based their life's work on publications within a university context. Edwards had performed his national military service with the Royal Canadian Air Force as a Pilot officer, training as a Navigator. He was a qualified flight crew license holder, and furthermore, he had an engineer's grasp of airframes and avionics. He was a member of the Royal Aeronautical Society, a member of the Royal Institute of Navigation and lastly was a London Liveryman of the Guild of Pilots and Air Navigators (now known as the Honourable Company of Air Pilots). All of these aforementioned professional bodies only admit persons who meet the professional criteria for membership, showing Edwards was well established in both Industry and Academia.

6.3 SHELL Model

After the publication of the Staines crash report, Dr. Elwyn Edwards wrote a paper and presented his new model at the British Airlines Pilots Association, *Outlook for safety. Man and Machine: systems for safety.* This new model explained the interactions and complexities that surround significant events – known as the SHELL model (Figure 6.2).

Figure 6.2 illustrates the SHELL model and indicates that the human in the centre of the diagram, shown as 'Liveware' (Figure 6.2, centre box), has

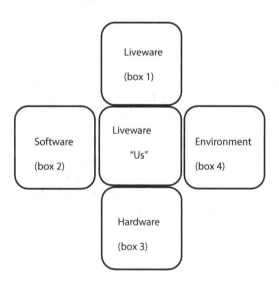

FIGURE 6.2
SHELL Model, Dr Elwyn Edwards (1972.)

various interactions with four other components. The principal assumption is that an accident is not a single event, but rather a series of events and interactions, and this new concept was used to explain the Staines crash.

Humans are considered complex, and we base our performance on multiple factors; thus, if sufficient negative events take place, a significant negative outcome would be possible in an aviation context. Basically, as a professional flight crew member (in the centre box), they will have interactions with Software, Hardware, Environment and Liveware. A brief description of each of these SHELL elements is as follows:

Liveware (box 1) represents the dynamic interactions with the subject to other staff in the context of the company. In an aviation context, this would include other pilots, ground staff, engineers, cabin crew, etc. Clearly, the Keys Liveware-Liveware links with other colleagues were ineffective, as demonstrated with the violent argument witnessed by other employees of the BEA. Additional evidence of such a breakdown is the graffiti found in both the crashed aircraft and operational aircraft, where fellow staff felt it acceptable to write strong-worded language detrimental to Key's character, such was the depth of feeling.

Software (box 2, Figure 6.2) represents the rules, requirements and regulations that are present in any employment. This would include company procedures; navigation data, including charts; policies, etc. In summary, Software is the non-physical aspects that define how activities and systems are organised. An underlying factor for this accident would be the noise abatement procedure (to retard the power levers as the aircraft passes over the perimeter fence). Another factor would be Key's strong dislike of strikes, along with him using his rank and seniority to force his beliefs on others.

Hardware (box 3, Figure 6.2) represents the physical aspects of aviation, such as type of aircraft; the flight controls (including the ergonomics); all the machine systems fitting within the aircraft, etc. A limitation of the Trident aircraft is the two separate levers that control the leading edge and trailing edge flaps, as both levers are required to be operated separated after take-off at the correct time, to ensure the aircraft performs correctly. It could also be argued that the ergonomic design of the lever in question is flawed. Post-Second World War, the end of the landing gear lever was redesigned to resemble a wheel. When selecting landing gear up or down, the pilot would feel for the distinct shape. The levers for the edge flaps also resembled wheels - clearly this similar experience was not considered by the Original Aircraft Manufacturer at the time of certification.

Environment (box 4, Figure 6.2) explains the physical conditions in which pilots work, such as temperature, humidity, pressurisation, noise, vibrations, etc. Additionally, political and social impact variables could also be included to represent the traumatic changes in society that have a bearing.

In addition to the four S, H, E, L components, the interactions between the centre box 'Us' and each of the four was proposed by Edwards as a boundary condition, and these boundaries and components are useful to explain human complex aviation accidents. Edwards's SHELL model was very appropriate for identifying the long list of contribution factors that were highlighted as causes in both the AIB findings and the subsequent government inquiry.

The SHELL model is very useful for analysing factors in complex aircraft accidents, providing explanations of how multi-layered interactions come together to form a complex failure. However, the model does not provide mitigation - such as a safety net that would allow a member of the flight crew to prevent the next Staines accident. As with all models, new versions were subsequently developed, with these allowing for alternate evaluations and a means of evaluating mitigation.

6.4 *The Impossible Accident* – Tenerife, 1977

Much of the worlds' aircraft fleet insurance is underwritten in some form through the Lloyds of London (UK) insurance market. In the early 1970s, the insurance market recognised the aviation risks and as a matter of course, considered worst-case accidents. The introduction of the Boeing B747 aircraft resulted in a new aircraft that can carry up to 550 passengers in a single 'economy' class configuration. The Insurance Underwriters in the Lloyds of London insurance market considered (prior to 1977) that the worst possible aviation accident scenario imaginable would be two B747s colliding with a total loss of the aircraft and passengers – and thus it was considered an *impossible accident*. Unfortunately, as history will demonstrate, the impossible worst-case events imaginable can occur.

The Tenerife accident occurred on 27 March 1977 on the Spanish Island of Tenerife in the Atlantic Ocean. Two passenger B747s collided at Los Rodeos airport (now known as Tenerife North Airport), one aircraft was KLM flight number 4805 and the other Pan Am flight number 1736. In total, 583 persons perished in the world's worst accident disaster, with a mere 61 passengers surviving the ground impact in the Pan Am aircraft. There were no survivors in the KLM aircraft. The view of the airport approach in recent times is shown as Figure 6.3.

The general background to this accident is interesting, as the events differ significantly from the Staines accident and thus the SHELL model is not as useful in terms of analysis.

In the early afternoon of 27 March, a terrorist organisation seeking independence planted a bomb at the main islands' airport, at the time known

FIGURE 6.3
Tenerife Airport Los Rodeos Airport on approach. (Ismael Jorda.)

as Las Palmas Airport (now Gran Canaria International Airport) which caused it to close immediately. All incoming flights were diverted from the Las Palmas, as per standard security protocols. Los Rodeos airport is located on the same island of Tenerife, so it seemed like a good diversion airfield. Unfortunately, Los Rodeos was a much smaller airport than the main airport (see Figure 6.3), and as such was not prepared for the high levels of traffic, having had very little preparation time further to the security bomb alert. Los Rodeos airport had no ground radar coverage, and under good weather conditions that would not necessarily be an issue, except on the day in question thick fog was moving in patches, which impeded visibility. There were lots of diverted aircraft at this small airport on the day in question, and the arrivals of large B747s complicated the ground movements significantly as other aircraft were obstructed by the much larger B747 aircraft.

The KLM aircraft that was at Los Rodeos airport was delayed, in part due to a lengthy refuelling activity. Additionally, KLM as an airline had recently introduced new directives for their pilots to limit their duty times, to prevent pilots from operating too long and being fatigued. The KLM flight crews were keen to load fuel, passengers, and take-off. The complication was the crews recognised that their duty time was going to be exceeded if they failed to take-off. The KLM P1 Captain, Veldhuyzen van Zanten, was a very senior member of the KLM flight operations staff, holding the rank of Chief Flight

Instructor, and spent much of his duties in the flight simulators for KLM, conducting flight assessments. These flight simulation assessments in practice meant that Zanten was controlling the flight simulators for his students, in addition to playing the role of Air Traffic Control (ATC). What is relevant here is Zanten's in-simulator use of non-standard phraseology and take-off procedures. The question that arises from this non-standard phraseology and take-off procedure is: who in the airline is correcting the Chief Flight Instructor's deviation from agreed practice? The KLM First Officer and Flight Engineer were also very experienced professionals with significant total recorded flight times, being in their 40s. The First Officer had only 95 recorded type hours on the B747 variant, but this was offset with 9,200 total flight hours.

The Pan Am flight crew had changed at Los Rodeos (unlike the KLM crews), and the Pan Am crew, comprising the Captain, First Officer and Flight Engineer were equally experienced in terms of total flight hours and type hours.

Another significant factor in this event was the departure of both aircraft from the stand, as illustrated in the simplified Figure 6.4.

The Pan Am flight (Figure 6.4 – Diamond shaped icon) was unable to depart off stand because the apron was full of other parked aircraft, and the KLM aircraft (Figure 6.4 – Triangle shape icon) was delayed by the fuelling activity. When KLM departed from the stand, the KLM taxied onto the active runway for the full length, and performed a 180° turn, in order to use the whole runway for the subsequent take-off. As the KLM aircraft departed onto the runway to position for departure, the Pan Am flight followed the KLM lead aircraft. ATC instructed Pan Am to depart from the runway when they reached exit 3, and then to continue via the taxiway, thus allowing the KLM aircraft to depart unimpeded. The weather was poor on the day in question, with lots of thick ground fog and cloud coverage, and coupled with this ATC had no ground radar to monitor the aircraft or vehicle movements.

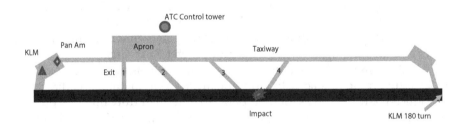

FIGURE 6.4
Simplified map of Los Rodeos airport showing the aircraft and the approximate point of impact on the runway.

The Pan Am aircraft, while taxiing on the active runway, missed the difficult 148° turn off the runway for exit 3 (due to impeded visibility), and attempted to remedy the ground navigation failure by seeking to depart the active runway at the next exit, 4. The KLM aircraft was already at the end of the runway, ready to start the take-off roll. The heavily reduced visibility due to fog continued, and neither of the aircraft could see one another. ATC was relying exclusively on giving radio instructions for all ground movements.

The KLM aircraft, keen to depart, commenced the take-off roll. Communication between ATC and aircraft had been unclear. ATCs spoken English was poor and the instructions to the aircraft contained non-standard commands. The First Officer of the KLM flight queried the take-off clearance, and ATC responded with the climb routing information, but included the term *'after take-off....'* in the instructions. The KLM aircraft started the take-off roll and, during the read-back of instructions, included the words *'we are now at take-off'*. ATC responded with *'Standby for take-off, I will call you.'* However, due to heterodyne, a phenomenon where two radio stations transmit on the same frequency simultaneously, the KLM aircraft were unable to hear the other station's transmissions.

While the KLM aircraft started the take-off roll, the Pan Am aircraft desperately tried to radio the tower to inform all users they were still on the active runway. In the ensuing seconds, both the KLM and Pan Am crews realised the full scale of the impending danger, but as the KLM aircraft was taking-off the aircraft only had around 4 seconds to react, and attempted to rotate. Unfortunately, the KLM aircraft impacted the rear section of the Pan Am aircraft, killing the rear Pan Am passengers. The KLM aircraft then crashed to the ground, killing all occupants.

In total, 61 passengers and crew located at the front of the Pan Am aircraft were able to escape from the damaged plane, although the ensuing post-crash fire claimed many lives.

6.5 Error Chain Model

While Edward's SHELL model was useful in identifying the causal factors and their linked relevance post-accident, a different and more simplistic model started to emerge in the mid-1970s. The new model is known as the Error chain and, in short, explains that a significant event is likely to transpire when seven (plus or minus two) events occur. This new hypothesis was supported by Boeing Commercial Aircraft Corporation's suggestion that an accident is unlikely to result from a single point of failure, such as the loss of a single hydraulic circuit. Rather, a series of events transpire, and when these

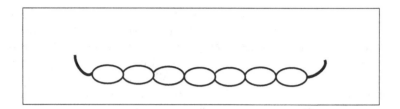

FIGURE 6.5
Error chain illustrating the individual 'links' of the chain forming an accident.

occur in a given flight the outcome is serious, with a crash or serious accident taking place.

Figure 6.5 represents the error chain model. The model is based on the theory that each event is based on 7 ± 2 individual events i.e. a minimum of five or maximum of nine separate events.

These chain elements are based on the following terms:

1. **Ambiguity.** Any time when two or more sources or persons do not agree.
2. **Fixation** or **Preoccupation.** The (excessive) focus of attention on any one item of event to the exclusion of all others.
3. **Confusion.** A sense of uncertainty, anxiety in the context of the event.
4. **No One Flying the Aircraft.** Lack of monitoring the current state of progress of the flight.
5. **No One Looking Out of the Window.** There is a 'heads-down' attitude, and flight crews fail to continuously look outside, to prevent events such as CFIT/mid-air collisions, etc.
6. **Use of an Undocumented Procedure.** This is the application of an unauthorised procedure in abnormal or emergency conditions – i.e. - deviation from the manuals and checklists.
7. **Violating Limitations or Minimum Operating Standards.** Deliberate intentional deviation from minimum standards, such as weather, landing or take-off speeds, etc.
8. **Unresolved Discrepancies.** The flight crews fail to resolve conflicts of opinion, information or changes in conditions that are necessary to generate consensus.
9. **Failure to Meet Targets.** Flight crews fail to achieve given targets, such as airspeeds, minimas, altitudes, etc.
10. **Departure from Standard Operating Procedures.** Flight crews intentionally depart from the prescribed SOPs, with the intention of saving time.

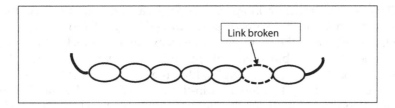

FIGURE 6.6
Error chain model with a LINK broken – resulting in 'no accident' concept.

11. **Incomplete Communications.** Communication is not fully effective because information is withheld. This failure to share information impedes other activities leading to misunderstanding, confusion, or disagreements.

When reviewing accidents, applying the Error chain model allows the reviewer to identify the major causal factors clearly and simplistically, with an understanding that the combination of 7 (±2) of the above factors will summarise the overall event. Another advantage of the Error chain is preventative philosophy, that if one single *link* event is removed from the equation, then the final result (i.e. the accident) will not transpire. This *break the chain* to prevent an accident philosophy is the principle advantage of this newer model, illustrated by Figure 6.6.

6.6 Flight Crew Training to Prevent Events

After Tenerife and the Staines accident, the aviation industry took bold steps to attempt to educate flight crews of the potential problems that can arise in an operational context, with the introduction of regular bespoke awareness training. The earlier recurrent training was known as Crew Coordination Concept (CCC), focused on identifying the elements that the Staines crash suffered from. Heavy emphasis was given to designated pilot duties on the flight-deck; the designation of PF exclusively to fly the aircraft; improved exchange of information between the crews, including transparent communication, and the monitoring and supportive role of the PNF. While the leading edge and trailing edge controls became integrated into a single control lever, the underlying problem of cohesion in the flight deck still continued to be an issue. This was in-part attributed to the different life experiences of the pilots, emerging either from a military background where *orders are followed and not questioned* or a civil training side where a more liberal approach to learning and human performance was taught. After the events surrounding the Tenerife crash, the training philosophy was adjusted to allow for the more updated Error chain model, succeeding the SHELL model.

In the early 1980s, the CCC training was expanded to include the professional roles that the cabin attendants played in an event. For example, Flight Saudia 163 on 19 August 1980 highlighted the importance of the roles of the cabin crews. Flight 163 suffered an in-flight fire shortly after take-off from Riyadh Airport to Jeddah. While Flight 163 was able to return back to make an emergency landing at Riyadh, when the aircraft came to a stop the cabin crews did not commence a full evacuation at this time. All 301 souls perished on this flight due to smoke inhalation from the cargo fire, resulting in Saudi Arabia's worst aviation loss of life (see Figure 6.7).

The CCC training was expanded to include the positive views and actions of the Cabin Crews and the training became known as Crew Resource Management (CRM). The inclusion of the cabin crews in flight crew training was very positive. There existed an *us and them* culture, where the flight deck door was known to be impede the sharing of information. This psychological barrier (the flight deck door) was mitigated by a new workplace communication strategy: the encouraged sharing of experiences of all the crews from all areas of the aircraft. Another useful improvement in this cooperation was seen in the airline's briefing rooms, where pilots and cabin crews brief separately, which has always been the industry norm. The solution to the 'them and us' culture was for the senior Captain of the flight to visit the Cabin Crew's staff briefing prior to departure. It was not uncommon prior to the 1980s for the Pilots and Cabin Crews to meet for the first time inside the

FIGURE 6.7
Saudi Flight 163 in-flight fire disaster, landing but all passengers and crews perishing before they could escape, 1980. (Leigh Kitto.)

aircraft, because many airlines had separate ground transportation for the crews (pilots or cabin crews) to move the groups from the security screening to the aircraft. The advantages of CRM and its effectiveness were demonstrated with other accidents transpiring in the later years of the 1980s. Weak CRM as a factor for the loss of life was demonstrated in other events such as the British Airtours (Manchester Airport, UK) fire in August 1985, Kegworth (East Midlands Airport) British Midland crash in January 1989 and the Air Ontario crash (Dryden Airport, Canada) in March 1989. All three of these accidents were used as case studies for airlines throughout the world, so lessons could be learned in the hope of preventing similar events.

6.7 Professor James Reasons' *Swiss Cheese Model*

Professor James Reason is one of the most important and revolutionary influences on error models and accident prevention in recent times. Reason, appointed as a Chair in Psychology at University of Manchester, UK, conceptualised a radically different model in the late 1980s, with numerous underlying observations in the *human condition*. Reason reviewed several well-known disasters (e.g. Piper Alpha Oil platform disaster, Challenger Shuttle Explosion, Chernobyl nuclear disaster, etc.) with a different view to most accident reports. The underlying principle that Reason surmised is that employees have *good* well-meaning intentions at work: deliberate acts such as sabotage are possible, but those deliberate, unlawful are already managed with the context of the law. However, events do take place in the workplace that are outside the direct control of the individual and the final outcome of the *mishap* is not intended. Reason identifies that safety critical industries, such as the nuclear power generation or space industry, are very rule-based and procedure-driven, yet both of these industries suffered very significant failures in the 1980s. Reason further stated that different types of errors (or single failures) existed, and these differed significantly.

An **Active Failure** is an occurrence where the operator of the system immediately recognises the failure: typically, these active failures are easy to identify due to their immediate identifiable nature (e.g. a mechanical pump failing).

The other type is known as a **Latent Failure**, alluding to the position that an error has been introduced into a system through an unintended action. The failure is hidden from view, and will only be detected at a later time, when other events transpire. Due to the latent condition, these failures are by their very nature complex and extremely difficult to identify in advance of a major event. Typically, latent failures can lie dormant for many years, and there is a likelihood of more than one event being associated with the system.

Reason suggests that different risk-averse industries have attempted to prevent failures by introducing preventative barriers. In the context of aviation, these barriers could include the following elements:

- The aircraft's initial design and manufacture is subject to 'a very careful oversight' by the National Authority certifying Aviation
- Pilots require extensive basic and type training on aircraft and must perform to a high standard in written and practical examinations
- Aircraft Engineers likewise require extensive basic and type training on aircraft and must perform to a high standard in written and practical examinations
- All maintenance to be carried out by Engineers who hold appropriate licenses (above), following written maintenance instructions from the aircraft manufacturer precisely
- All calibrated specialist tooling is in good condition, checked and certified at given intervals
- The planning of all maintenance activities is performed by technical staff, who review the supplied literature from the manufacturer and supply it in a working format to the engineering staff
- Etc

Each of the above bullet points is considered as a *barrier* of some protective layer, because historical events have taken place. These *barriers* will require some 'tuning' to prevent the same event happening again. It is this breaching of *barriers* or protective layers that Reason explains as the Swiss Cheese model. Reason's book, *Human Error*, 1990, explores this new Swiss Cheese theory, and his subsequent book *Managing the Risks of Organizational Accidents* 1997, contains numerous airline observations.

Figure 6.8 represents Reasons' Swiss Cheese model. The layers of Swiss cheese with holes in them represent the barriers that have been introduced. All barriers have 'holes' that appear and go away and are considered dynamic. Errors can penetrate single layers of protection, sometimes through multiple layers of defence. If a single barrier rejects (i.e. identifies the error as a problem), the protection task has been successful, and the potentially tragic outcome averted.

No single system can be 100% safe all the time and prevent all accidents/ serious events. When the errors do occur, which is expected, if the error is able to penetrate through the *barriers holes* it will pass to the next layer. If the error is of a type that can penetrate through all the barriers, then an accident or significant event will occur. In an aviation context, this can include the loss

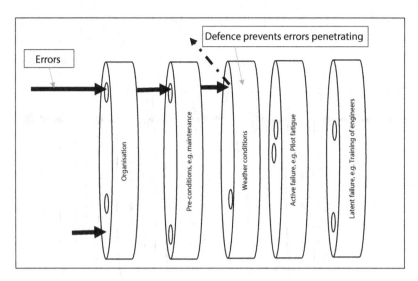

FIGURE 6.8
Swiss Cheese model showing individual barriers intact.

of the aircraft, the passenger's lives and any other persons on the ground that interact with the respective aircraft. Figure 6.9 illustrates how these barriers can be identified and how the error(s) can then penetrate through all the layers of protections resulting in the accident.

The Swiss Cheese model explains many events, and the significant difference between this and other models is the awareness factor that can be gained by educated individuals on the front line i.e. pilots, engineers, Air Traffic controllers, etc. Additionally, the second-tier staff members who are not directly involved in hands-on maintenance or flying activities, but more in support roles (e.g. technical service general engineering staff), can be made aware of the possibilities of their own creation of events that, if undiscovered, will be classed as latent failures.

Reason also makes a profound observation, namely the staff at work go to perform their duties to the best of their ability. Sometimes things do not go as planned, and when the management of the given organisation discovers this, the management feels it necessary to impose punitive measures. Reason states that typically, the employee with the defective performance is sanctioned and privileges are often withheld (often with a financial implication): the individual is then subjected to a retraining activity because it is believed they require upskilling to better perform. Neither of these actions are effective, and the blame and train culture and activities are counter-productive and ineffective, as a future failure is probable. If the initial failure occurrence was not a deliberate act, then why would a previous sanction and additional training help prevent such possible future failures? Furthermore, when the

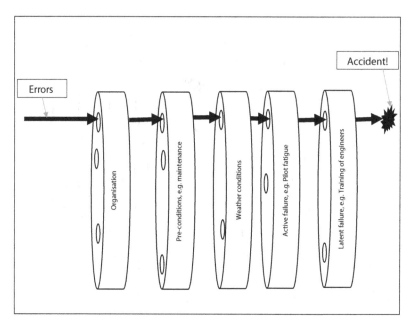

FIGURE 6.9
Swiss Cheese model showing all barriers breached and the accident.

next failures do occur, management takes the view that the individual has chosen to fail; therefore a more severe punishment is required to remedy this lack of performance. Reason rightly explains (1997) that staff in such an organisation will be reluctant to report failures, events, and occurrences to the company, and the likelihood of more latent failures becoming undetected is highly probable. Staff will cover up their own failures, and the management will be totally unaware of the levels of failure. To improve upon this negative, vicious circle, Reason (1997) promotes the substitution test, namely for the manager to put themselves in the employee's shoes. The substitution test simply states, *if you find you could make this mistake, what are you (the manager) going to do about it, and if you state you wouldn't make the mistake, then what makes you so special!*

Reason's application of the Swiss Cheese model to commercial aviation in the late 1990s experienced incredible levels of success and adoption. This was because the worldwide accident rate at this time suggested that if nothing was put in place by 2015, a hull loss and significant loss of life would be happening on a weekly basis. The adoption of Swiss Cheese theory and the mandated training of HF for all staff in an airline who can have an operationally detrimental effect, as they produce latent and active errors, has been instrumental in improving the safety record of the industry since 2001. This mandated training requirement has resulted in a significant reduction in the losses of aircraft and reduction in the loss of life, as supported by data published by Boeing and various aviation safety organisations. Prior to the mandating of initial and recurrent HF, the UK Civil Aviation Authority (CAA) conducted industry briefings for all accountable staff in UK registered airlines (between 1999 and 2001). The CAA's message to the airlines was the necessary improvements in human performance and accident reduction rate should not be 'sold to staff' that the principle inclusion of this subject will save lives. This is because the problem with measuring safety and accidents is a simple dilemma: commercial aircraft accidents are very infrequent. The lack of accidents leads to a false sense of achievement because in the preceding month, months, years, etc., your organisations' lack of events implies that the airline is fully resilient to possible future crashes. Reason confirms this difficulty, as demonstrated with the lack of catastrophic experiences in other non-aviation industries followed by massive losses (e.g. Challenger, Chernobyl, etc.). The CAA stated in presentations that the corporate commitment will be present when safety improvements result in fewer technical delays, fewer errors and a more productive work environment. Furthermore, the *modus operandi* suggested to gain the necessary full corporate commitment from the airline was to imply that '3% savings can be achieved from the engineering budget'.

Therefore, the financial savings from performing the work correctly, first time and every time has a significant, measurable cost factor, and thus the accident reduction was a welcome by-product of better working practices.

6.8 Safety Management Systems

The SMS is a collective system of elements with the principle objective of enhancing safety. This is achieved through an organisation creating their own bespoke practices and operations from four distinct areas:

- Safety Policy - the structure of how the organisation responds to legal requirements and operational deliverance. This policy creation and commitment requires corporate endorsement and defines methods/processes and objectives.
- Risk Management - in the most simplistic evaluation, individual risks are assessed against likelihood and severity. For example, the likelihood might scale from *highly improbable* through to *highly likely*. Likewise, the consequence might be scaled from very minor damage or injury (e.g. a bruise) to catastrophic event (e.g. crash with loss of life). Using this simplistic two-parameter assessment, a matrix, or number matrix (often colour-coded), can be produced. This contains the multiplication of

factors in respective centre boxes from the two corresponding elements. The highest numbers (e.g. certainty of event, death being the result) will score the largest risk value. The use of this model is to quantify the risk. Control measures to mitigate the risk of occurrence and effects are included.

- Safety Assurance evaluates independently how well the system is performing and whether the risk management is effective. Measurements can be performed to assess losses in a unit, either financially, loss of technical dispatch reliability (a commercial aircraft being available for revenue service) or even sickness and absenteeism.

- Safety Promotion is the continuous training and communication activities for all staff, to enhance the safety culture in the whole business.

The SMS is the combination of all four of the above bullet point items. While the content may appear broad and uninspiring (to some), the implication in the aviation sector is immense. Firstly, the SMS has been mandated in commercial aviation since 2013 by International Civil Aviation Organization (ICAO), and although ICAO has no legal authority, the regional aviation authorities (including the European Aviation Safety Agency) can and do have the legal framework to ensure corporate organisations fully participate. In the context of aviation, an interesting inclusion is the extension of the SMS to third-party repair and overhaul stations, because it is not only airlines that experience problems pertaining to safety that require close management and improvement.

An SMS is not a transferable product or software solution that can be sold from one company to another. Every organisation is different because staff in companies have different working practices and procedures. The SMS is, by definition, a unique collection of tools that have been adapted and applied to an organisation.

Other industries (aside from the aviation sector) have adopted the SMS as a regulatory minimum. The maritime industry has adopted the practices, as have certain national railway authorities, and the resemblance to the successful aviation adoption is very noticeable. The guiding principles are identical, namely deliberate acts (i.e. sabotage) require deliberate sanctions (i.e. custodial sentences), but as most deviations from approved procedures or processes are because of external factors outside of the control of the individual, the substitution test of the management (post an event) demonstrates the effectiveness of human performance and SMS.

6.9 Conclusions

The aviation industry has demonstrated for many years the willingness to learn from mistakes and events. After a significant event (such as a plane crash), the causal factors are identified and the national accident investigating bodies make these findings public via their published reports. The openness of this attitude has allowed for a number of models to be developed over the years, with each model explaining the different inter actions and complexities that appear. The models have all been developed and applied to different events, with the principle objective to learn from the past, because there is no advantage to the industry if the same type of accident occurs over and over.

While the models are all excellent in explaining factors pertaining to previous accidents, the Swiss Cheese Model is particularly advantageous in allowing an organisation to benchmark *how close to the edge* it currently is, without the need to experience catastrophic events. This self-awareness is the underlying principle that aligns many existing procedures and management processes to form an effect SMS.

The weakness of any safety critical system is the infrequency of the recorded events, and senior management must continue to ensure they fully appreciate all the challenges that the staff experience on the front line (e.g. pilots, engineers, air traffic controllers, airline engineering planners, etc.). Failure of the management to grasp the actual challenges and control them will likely result in Swiss Cheese 'holes' appearing in existing systems. History shows us that when enough of these holes line up, it is more likely that a catastrophic event takes place. The senior management (from the Chief Executive Officer down) must accept that as named 'accountable persons' they will be the first airline employees to be prosecuted by the national aviation authorities, thus the 'us and them' culture cannot be an excuse after an event.

The next chapter will discuss the changes to physical barriers and processes associated with security events in commercial aviation, such as 9/11. Additionally, other events including relevant non-aviation security occurrences are included, such as liquid improvised explosive ordinance.

7

Aircraft Security

7.1 Introduction

This chapter will discuss the development of the aircraft's flight deck separation between the pilots, the cabin crews and the passengers, and the implications of the physical barriers. The development of aircraft is considered with the introduction of flight deck doors being an improvement from a curtain. Even though the doors were unlocked and freely accessible to all occupants, the presence of the door encouraged a division of attitudes. Forward of the door were the pilots who flew the plane and made the critical decisions; aft of the door was the cabin crews who looked after the passengers and received orders from the pilots. This thinking is partly due to regulations that have stemmed from the nautical industry, which aviation has developed from in a legal transport context. Like ships, the Captain is fully in-charge of the aircraft, the crews and the passengers.

A number of accidents/events occurred in the 1980s and 1990s relating to the crews (pilot, cabin crew) not communicating effectively, which was partly attributed to the flight deck door being kept closed. Post the 9/11 terrorist attacks on US soil, the flight deck door became a reinforced structural item, being bullet-resistant and locked to keep out would-be aggressors. Additionally, the use of Sky Marshals on flights worldwide became more pronounced in order to covertly manage cabin security and prevent terrorists gaining access to the cabin. Lastly, hydrogen peroxide-based explosives have been extensively used in ground-based terrorist attacks from 2005 onwards, with potential attacks on passenger aircraft in summer 2006 being foiled by UK Security Services. Immediate restrictions on hand luggage liquids were instigated worldwide as an effective counter to this new method of downing commercial aircraft by suicide bombers.

7.2 Flight Decks with Curtains and Doors

The mass of the aircraft is a significant factor in aviation, due to the physical performance, including the lift generated. The engines provide the thrust that propels the aircraft forward, which in turn causes the air forward of the aircraft to flow over the wings, and the wings deflect the air flow around them, thus creating lift.

The weight, or more correctly, the mass of the aircraft must be overcome by the lift generated by wings, so if the aircraft is unable to reach sufficient forward speed to create sufficient lift, the plane does not take off. It is this mass-limiting balance that aircraft designers have addressed, even from the earlier days of flight. Early powered aircraft used technologies developed in fabric-covered ultra-lightweight rowing boats, coupled with a motorcycle engine, to allow for powered flight. Typically, the militarisation of new aircraft technology has allowed these small aircraft to develop further, to fly faster, higher and to carry more mass (e.g. fuel, bombs, passengers, etc.). All of the aforementioned mass limitations have been addressed effectively since the end of the First World War, and the ongoing development of high bypass turbine engines has resulted in much greater engine performances.

These early pre-war aircraft had no physical separation between the flight deck and the passenger cabin, partly due to mass limitations, because additional interior structures reduce the carried mass of the aircraft. When a partition was introduced between the flight deck and the cabins, the 'lightweight' solution was a curtain, yet this did not provide the necessary levels of privacy and noise isolation for the flight deck occupants. The advancement of plastics, resins and composite structures in the late 1950s and early 60s allowed for a newer lightweight aircraft interior to be used extensively throughout the interior fittings. With the dedicated privacy of the flight deck, created from a walled partition, the physical presence has also created changes in behaviours.

As discussed in Chapter 6, the culture of *us* and *them* between the pilots and cabin crews has been a detrimental factor for many years. The physical presence of the separating bulkhead and door has played an active part in many events, in particular the British Midland Flight 92 accident on 8 January 1989. The B737-400 was being flown from Heathrow to Belfast Airport (Northern Ireland). In-flight, the flight crew misdiagnosed engine vibrations as a pending failure of the starboard (right side) engine. The pilots turned off the starboard engine and the aircraft attempted to land at the maintenance base at East Midlands airport. Prior to landing, the flight crew made a public address to the passengers and cabin crew that the reason for the diversion was because of problems with the engine on the right. The passengers on the left side of the aircraft could see burning debris being ejected from the exhaust of the so called 'good' left engine. No one from the passenger cabin (be it cabin crews or passengers) challenged the pilots' PA. No one brought the error to the attention of the pilots that the passengers and cabin crews could see that the left engine was the defective engine. On approach to East Midlands Airport, the left engine began to deteriorate further and the engine began to fail, reducing the thrust output. The pilots were unable to restart the right engine in time, and the aircraft stalled short of the runway, crashing into the embankment of the M1 motorway just south of junction 24. There were 47 fatalities in this crash, 74 serious injuries out of the 126 passengers and crew on board. The absence of a post-crash fire enabled the injured passengers and crews to be removed from the aircraft without further injury. After the publication of the UK Air Accident Investigation Branch report, the facts were established and the aviation industry tried to *learn lessons* from this event. They did this by educating pilots and cabin crews of the need to actively listen to PAs, and if a mistake is made, such as in the case of the British Midland accident, for the crews to feel comfortable to openly share those views by communicating with the Flight Deck. UK airlines used this event as a case study for their crews to discuss, highlighting the importance for a cabin crew member to feel empowered enough to enter the Flight Deck and convey their concerns. Likewise, some airlines encouraged their pilots to consider leaving the Flight Deck door open, to bridge this *us and them* culture, and to remedy the communication difficulties that have occurred over the years.

7.3 British Airways 2069, 29th December 2000

This British Airways flight was departing London Gatwick airport, bound for Jomo Keyatta Airport in Nairobi, Kenya. The aircraft in question (see Figure 7.1), a B747-400 aircraft, was fully loaded with 379 passengers and 19 crews, jetting off to Kenya to see in the New Year.

British Airways had worked hard over the years on crew resource management, more so since the British Midland crash, to encourage good open communications between the pilots and the cabin crew. On this flight, like so many other flights, the crews worked well together with the cabin crew visiting the flight deck regularly, and the flight deck door was opened/closed without any fuss in this normal working environment. Later in the flight's cruise all was normal, so the Captain left the flight deck to take a rest, leaving the two other first officers at the controls. At around 5 am local time flying at 35,000ft, a mentally unstable Kenyan passenger, Paul Mukonyi, ran through the upper deck cabin, into the flight deck and attacked one of the first officers. In a very violent struggle between Mukonyi and the pilot, the autopilot was

FIGURE 7.1
British Airways 2069 aircraft involved in the cabin incident (passenger attempted homicide/ suicide). (Paul Link.)

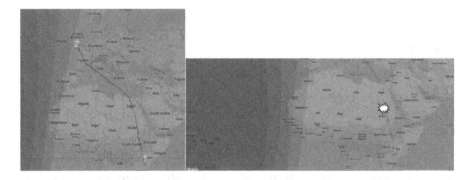

FIGURE 7.2
BA 2069 Flight track indicating the position of aircraft during the storming of flight deck by perpetrator. (Google Maps.)

accidently disengaged and the plane rapidly descended, with a maximum decent rate of 30,000 ft per minute. The passengers, Captain and crews managed to subdue the assailant, and the aircraft continued on to its destination. The approximate location of the aircraft on the planned flight track during the event is shown in Figure 7.2. In the subsequent investigation, a number of reports after this event suggested that Mukonyi wished to take the aircraft by force, or commit a multiple homicide/suicide event. Although the aircraft later landed safely in Nairobi as scheduled, a small number of passengers required additional medical treatments due to the violent manoeuvres that the aircraft performed. The reporting of this event was very high profile at the time, as the passengers included Bryan Ferry, the musician and his family, the Goldsmith family and Jemima Khan (nee Goldsmith) who married the ex-Pakistan international Cricketer Imran Khan, currently the Prime Minister of Pakistan at the time of writing. The shocking event was captured on a passenger's video camera as Mukonyi was dragged from the flight deck by passengers and crew.

Unfortunately for British Airways, this was not the sole incident where a passenger had tried to *storm* the flight deck. Two years earlier on 13 February 1998, the Stone Roses lead singer, Ian Brown, was flying on BA 1611 from Paris to Manchester. In-flight there was a significant event where Brown threatened to chop off the flight attendants' hands. Later Brown attempted to storm into the flight deck during the final descent phase of the flight. Brown was arrested on arrival, prosecuted and was sentenced to a custodial detention of four months in a UK prison.

These events have highlighted the ease at which an individual could gain access to the aircraft flight deck, compromising the safety of the crews and the aircraft. The physical barriers were kept to an absolute minimum between the passenger cabin and the flight deck, yet this breech of security, even with the extensive media attention, did not change the working practices at the time.

Aviation security is a concept and skill that has been taught to all pilots and cabin crews for many years. The syllabus includes events that rarely take place, such as hijackings. More frequent events are roleplayed, including the safe management of intoxicated passengers and restraint techniques approved by the carrier, including the use of the handcuffs and restrains that are carried on all aircraft. In-flight security events (including the restraint/cuffing of passengers) have been a common occurrence for many years, partly attributed to the availability of alcohol onboard the aircraft.

One such event was experienced by an Asian carrier in the late 1990s, where a passenger on an ultra-long haul flight from Asia Pacific to Europe became aggressive and abusive to the cabin crews and passengers. The cabin attendants tried to manage the problem, but this had little effect on the drunk individual. The Captain decided to leave the flight deck and tell the passenger of the consequences of their actions. Being an ultra-long haul flight, the aircraft had a complement of 4 pilots. The Captain's strong words of advice were ignored, so assistance was sought from strong, able-bodied passengers: a number of professional international rugby players assisted the crews in physically subduing the intoxicated passenger, while the cabin crew cuffed him (with his hands in front of his torso).

The verbal abuse and disruptive behaviour continued in the passenger cabin, even though the passenger was seated and cuffed. Passengers nearby were distressed, so the cabin crew decided to move the passenger from the cabin to the toilet (with the passenger still cuffed), and to lock him in the toilet to prevent him from further interacting with the other passengers. While the decision to lock him in the toilet might seem like a reasonable solution, the reality was somewhat different. While in the locked toilet, the cuffed passenger decided to burn toilet papers and paper hand towels with a cigarette lighter that was concealed on his person. Fortunately, all aircraft toilets were fitted with a smoke detector that alerts crews throughout the aircraft if smoke and fire was detected.

The toilet door was removed, the cuffed occupant forcibly removed and the fire extinguished. The perpetrator was later arrested by the police upon landing and prosecuted for endangering the aircraft, the crews and the passengers, resulting in a lengthy custodial sentence.

The importance of this smoke detector is highlighted in a previous accident with Air Canada on 2 June 1983, where a toilet fire in-flight caused the Air Canada aircraft to land at Cincinnati Airport, with 23 fatalities. The danger of in-flight fire is a significant worry to any pilot, as the risk of the *flash over* phenomena could result in a non-survivable cabin environment. It would be almost impossible to extinguish a fire with the apparatus in the aircraft if this point has been reached. *Flash over* is achieved when the temperature of the fire raises the surround materials to a level that exceeds the flash point temperature, meaning that all nearby materials spontaneously combust without the need for a flame: everything flammable *bursts* into flames at this stage, including the cabin occupants.

These events provide a worthwhile lesson to all airlines and crews trying to manage a confrontational cabin passenger, which is not to use the toilets as a makeshift detention cell, as the risk of arson can lead to a more catastrophic event.

7.4 Year 2001 Events and Post 9-11 Modifications to the Flight Deck Door

2001 was the year that changed aviation throughout the world, owing to numerous acts of terrorism.

In the summer of 2001, the largest aviation insurance claim in the Lloyds of London (UK) insurance market took place. Most readers would expect this to be the 9-11 events, but it relates to the Bandaranaike Airport attack on 24 July 2001 (Figure 7.3, illustrating the close proximity of the Sri Lankan military complex to the passenger terminal building).

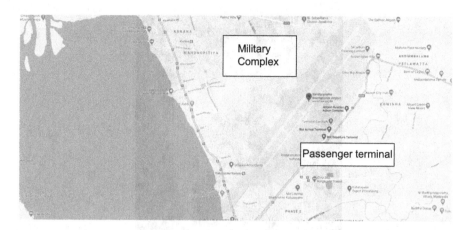

FIGURE 7.3
Map of Bandaranaike Airport, Sri Lanka, during terrorist attack 2001. (Google Maps.)

The Sri Lankan civil war had waged since 1983, with terrible violence and loss of life. In the early hours of 24 July 2001, a small number of Tamil 'Black Tiger' fighters entered the airport perimeter via the military complex. Using a variety of weapons, they destroyed 8 military aircraft on the ground (damaging 11 more), then crossed the runway to the civilian side of the airport. At the terminal, they continued to destroy a further three Sri Lankan Airlines Airbus aircraft and damaged two Airbuses, in addition to damaging the airport infrastructure. The insurgents were eventually killed by the military during the ensuing counterterrorism operation. Although the airport was closed for approximately 14 hours, further disruptions and negative effects on the tourism industry had a measurable impact on Sri Lanka's economy and GDP. This event focused the Sri Lankan government on resolving the civil war, although a further 8 years of conflict took place before the Tamil Tigers accepted a lasting ceasefire.

The most infamous event of 2001 was the coordinated attacks on the USA by the foreign terrorist group al-Qaeda, using commercial aircraft. Four domestic flights (two from Boston's Logan airport, the others from Washington Dulles Airport and Newark Airport) were taken over by terrorists posing as passengers.

At the time of the events, security for domestic travel in the USA was more relaxed in comparison to international flights, in terms of screening of baggage, screening of passengers and hand luggage, etc. All four flights resulted in terrorists gaining control of the aircraft at approximately the same time, using the aircraft and occupants as a weapon to attack the Pentagon and both of the iconic Twin Towers of Manhattan, New York (see Figure 7.4).

Passengers on these flights were able to communicate with friends and families using satellite phones, explaining that the terrorists' hijacking

FIGURE 7.4
Post-apocalyptic image highlighting the destruction of the World Trade Centre on 11th September 2001.

method was the use of sharp metallic hand-carry items (e.g. scissors). Until this event, airline security training for hijack situations was to acquiesce to the perpetrators of the violence. This event demonstrated that some hijack events have no resolvable outcome when the acquisition of money or transport is not the goal. Rather, terror and maximising publicity for the attack is the sole purpose, with the principle objective being to kill as many souls as possible.

While governments have always had military aircraft on standby to intercept hostile aircraft, the interception and shooting down of a civilian commercial aircraft on domestic soil has never been recorded. Despite air traffic controllers being aware that multiple aircraft had been taken over, they were reduced to spectators of the events as they unfolded. As the records have indicated, the US Airforce's Quick Reaction Aircraft, capable of intercepting these four commercial aircraft, were not available on that day.

All four aircraft were deliberately crashed that day with no survivors from the aircraft. In the buildings that were attacked, large numbers of persons lost their lives in the world's worst suicide/homicide aviation event to date.

In the immediate aftermath of these four events, the Federal Aviation Administration grounded all flights temporarily. When flights did later resume, a much stricter cabin screening process was instigated worldwide, with the prohibition of all sharp objects carried by passengers or their hand luggage. Visible screening activities now take place at the airport as passengers and crews transition from *landside* to *airside*. All hand-carry luggage undergoes an X-ray examination, and passengers undergo a metal detection screening, with the possibility of a pat-down examination by security staff. Another immediate change was the removal of metal cutlery from the aircraft, with plastic utensils replacing the metallic items. Additionally, to mitigate another potential terrorist attack, respective government counter-terrorism units have increased Sky Marshal activities on commercial flights. While most of the operational details of these activities are not published, the principle of the service is to prevent passengers attempting hijack events. The Sky Marshals are armed with firearms, and their covert approach is to appear as a normal travelling person who will only intervene should it prove absolutely necessary. Occasionally, Sky Marshals have intervened in security events in the airports' *airside* environment, such as the killing of Rigoberto Alpizar, a passenger of American Airlines 924 (Miami International Airport) on 7 December 2005. Alpizar was shot and killed by Sky Marshals when Alpizar attempted to run through the passenger cabin to egress the aircraft in a frantic state. The Sky Marshals identified themselves, but when Alpizar implied he had a bomb in his backpack and ran up the airbridge into the terminal building, they opened fire.

The most significant modifications to commercial aircraft post-9/11 are the strengthening of the structure around the flight deck doors; the new door locks; and the new closed-door protocol. These requirements were announced in the autumn of 2001, and the ICAO mandated all airlines to complete these modifications by November 2003. The structure around the flight deck door has been strengthened to be resistant to sustained attack (Figure 7.5).

The door panels also contain ballistic-resistant Kevlar material in the composite, with the door locked at all times, i.e. when the aircraft is moving under its own power. These strengthened doors have resulted in the flight deck being highly secure from the inside, with it being unlikely to be forced open from the passenger cabin. Cabin crews can unlock the door from the outside under certain conditions, but this process is protracted and can be overridden from the inside by the flight deck occupants. The doors also have an internal lock to prevent entry from the cabin. In summary, once the flight deck door is closed, locked and deadlocked from within the flight deck, there is little possibility of opening the flight deck door. The implications of this will be explored in the next chapter regarding the GermanWings 9525 murder/suicide event on 24 March 2015.

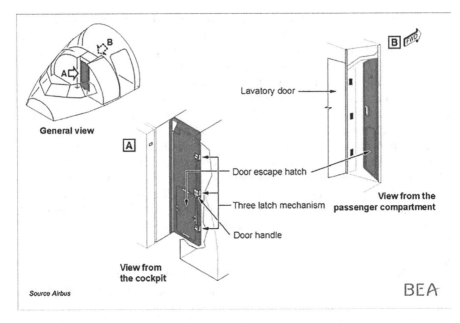

General view

Lavatory door

Door escape hatch

Three latch mechanism

View from the passenger compartment

Door handle

View from the cockpit

Source Airbus

BEA

FIGURE 7.5
Airbus A320 reinforced flight deck door. (BEA.)

On 7 July 2005, London witnessed one of its worst domestic terror attacks, performed on its ground transport infrastructure by Islamic extremists. Four bombs were detonated in London, three on underground tube trains and one on a double-decker bus. This became known as the 7/7 London terrorist attack. An identical iconic double-decker London bus is shown as Figure 7.6.

The 7/7 event killed 56 (including the four terrorists) and injured 784 innocent members of the public. This is relevant to aviation due to the choice of explosives used in the attacks. The explosive was derived from concentrated hydrogen peroxide (a liquid) to form a complex unstable chemical known as tricyclic acetone peroxide. While the discovery of this explosive is not new (discovered by Richard Wolffenstein in 1895, and further synthesised by Adolf von Bayer and Victor Villiger in 1899), raw materials for bomb production were readily available, as hydrogen peroxide has widespread use, including cleaning products and hair dyes. The availability of this potentially everyday explosive was known by the terrorists and has been exploited by terror cells throughout the world in the production of improvised explosive devices.

In the summer of 2006, the UK security services intercepted terror communications that implied an aviation terror (homicide/suicide) event was imminent. The *modus operandi* identified was for suicidal terrorists to carry liquid explosives *on their person* or in hand luggage onto a commercial aircraft, and

FIGURE 7.6
Red London double-decker bus, identical to the bus destroyed in the 7/7 terrorist attack.

to detonate these liquids in-flight, resulting in a total loss of life as well as the aircraft. All liquids entering the passenger cabin were banned on 10th August 2006, with many countries following the UK's initiatives soon after. The intercepted information suggested that baby food (that would be carried with a baby) would be replaced with this liquid explosive, indicating the extreme lengths that would-be attackers would stoop to achieve the downing of a commercial aircraft. Later, in early November of 2006, the UK government allowed up to 100 ml of liquids to be carried by passengers as part of their hand luggage, but these liquids were subjected to additional security screening.

The restriction on liquids in hand luggage continues at the time of writing, indicating the potential for liquid explosives to blow up a passenger aircraft.

7.5 Conclusions

The changes to aircraft operations post-9/11 have resulted in significant changes to working practices for staff and travel experiences for passengers. The locking of the flight deck door has resulted in pilots becoming even more isolated, with the cabin crews only able to visit the inside of the flight deck under very controlled circumstances. The encouragement of cabin crews in the 1980s and 90s to visit the flight deck and share their concerns has now been replaced by the concept of the flight deck as fortress. While the cabin crews still have the option to communicate via intercom, the loss of the *face to face* element *is the price paid* for this additional level of security and isolation.

Lastly, in the post-9/11 world, the aviation security strategy has moved significantly away from acquiescing to the demands of hijackers or terrorists, and instead to the widespread deployment of covert Sky Marshals aboard flights. This new worldwide strategy means that potential perpetrators can expect to encounter an armed deterrent in the passenger cabin, which has had a positive influence on the aviation transport sector.

The next chapter will address specific, unusual aircraft events - including negative aspects of the locked, secure flight deck - that have resulted in significant loss of life and aircraft.

8

Unusual Losses of Aircraft

8.1 Introduction

This chapter will identify several unusual aircraft crash events from the 1980s to present day that highlight the tragic consequences of murder/suicide related aircraft losses. The events that are presented have all been investigated by national accident investigation authorities, and the reports have been publicly published. Typically, each report is a very substantial document with hundreds of pages of factual content, in addition to appendices illustrating the background, facts, and the findings of the investigators. The discussion in this book is not intended to replace or supersede any such published report, rather the pertinent points are identified for the reader and placed in the context of aviation safety and security. Some of the accident investigations are incomplete, awaiting the discovery of further physical evidence, such as the MH370 event. For simplicity, this event will be included as a possible murder/suicide occurrence, due to the weight of evidence regarding its flight characteristics as well as other evidence that has arisen over the years.

Lastly, current aircraft weaknesses are included in this chapter to highlight the real potential for a catastrophic type event. The prediction of such events is always continuous, but is based on the combination of technical performance of the aircraft, historical events and known operational weaknesses.

8.2 Korean Air Lines KL 007 Shoot Down, 1 September 1983

Korea Airline KE 007 was a scheduled flight from New York's JFK International Airport to Seoul Gimpo Airport. The aircraft was a Boeing B747-200 series passenger aircraft, and on the 1 September 1983 this flight was shot down with 269 souls on board.

FIGURE 8.1
Korea Airlines KE 007, 1 September 1983 Flight plan and actual track. (NTSB.)

The first leg of the flight was uneventful, from New York's JFK to Anchorage, Alaska. The aircraft refuelled in Anchorage, and took-off bound for South Korea, as illustrated by the solid black line in Figure 8.1.

During the extended flight over the North Pacific Ocean, the flight procedure required the flight crew to overfly the town of Bethel, Alaska (very close to the Pacific west coast) and enter the North Pacific (NORPAC) flight routes that link the Alaskan and Japanese land territories. Normally aircraft overfly land-based VHF Omni Range (VOR) radio beacons, and within a modern aircraft the beacon is identified with the respective bearing (`To' or `From') and the distance from the aircraft to the beacon is shown. However, as it is difficult to operate such beacons on floating structures, a different technology is employed when travelling across large bodies of water, namely Inertial Navigation Systems (INS). Older INS technology used a spinning mass with accelerometers that recorded the rate of change in the three axes of direction, but the newer versions are based on ring laser gyros. These are very accurate once `aligned' as they use the laser's light-based interference patterns to determine the rate of change (acceleration).

The autopilot system that was fitted to this aircraft required the navigation system to be changed from the VOR radio data to the INS data inputs within a short distance (approximately 7 nautical miles) from the ground-based VOR station. While the INS system was functioning correctly, the flight crews did not select the correct data input source for the autopilot. Figure 8.1 illustrates the dashed line track that should have been flown with the correct procedures, and the solid line is the unintended flight track that took KE 007 into Russian airspace, which was prohibited at the time of the incident.

In the 1980s, the Cold war was ongoing and tensions between the West and the Soviet Bloc resulted in numerous small military flashpoints. On this flight, KE 007 entered Soviet airspace at 15.51 UTC. A compounding factor was that some of the Soviet's ground radar detection was not fully operational; the detection of this civilian aircraft came as a shock to the Soviet radar and ground commanders, because their long-range equipment was not working and the short-range equipment had made a late detection. As a result of this unexpected entry, the military scrambled interceptor aircraft to locate, identify and, if necessary, take action in order to repel the intruders. Unfortunately, due to the nature of interceptors and their limited fuel payload, range, etc., the military flight crews were unable to obtain a clear visual flight confirmation, even though they reported to their commander that they had done so. This was because the intercepted aircraft was dark, and the fighter pilots claimed they could only see the anti-collision beacons and strobe lights. The Soviet military ground commanders believed that the `intruder' was a military-type aircraft, and thus they gave the interceptors orders to shoot down the so-called hostile aircraft before it could leave Soviet airspace and escape. The Korean Air Jumbo changed altitude around this time, and the fighter aircraft flew past the aircraft. At no time did the interceptors or the Jumbo successfully make radio contact with one another prior to the following tragic events.

In the confusion of the fast-changing situation, with the interceptors running low on fuel and the eagerness of the ground commanders to be decisive, two air to air missiles were launched by the Soviet fighters at the Boeing B747. The Jumbo was fatally damaged, and while the aircraft did not explode or break up during the attack, the damage to the aircraft and flight control systems resulted in the Korean Air flight crew being unable to control the plane. The aircraft flew for a short time and descended to a low altitude before crashing into the sea close to the Moneron Island, a Russian territory. All aboard were killed. Attempts by the American military to locate KE 007 on the seabed were scuppered by the Soviet military, who denied knowledge of the crash or the location, and prevented the Americans from starting Search and Rescue (SAR) activities within Soviet waters.

Many years later, the Soviets admitted to sending SAR operations to the location of the crash site. More importantly, the Soviet military secretly recovered both the flight recorders but denied any knowledge of the crash location, nor the recovery of the CVR and FDR recorders. It was not until 1992, some nine years later, that these recorders were returned to South Korea as an act of good will by the Russian President, Boris Yeltsin. It was only with the return of these recorders that it was possible to download the actual recorded data for a full analysis. For many years, the West was not able to understand how a Jumbo could accidently fly into Soviet airspace unintentionally, with such tragic consequences.

After this event, the US government wanted to prevent such an event from happening again by allowing civilians free-use of the US military's Global Positioning System (GPS). Using the GPS system would allow civilian users to be able to identify where they were on the planet, by receiving a minimum of four separate satellite signals. Later, the aircraft manufacturers modified their aircraft to make it clearer to the pilots which system was being used with the autopilot system, be it ground-based navigation data or INS data.

This event highlights the effects of small procedural errors made by pilots that can take their planned flight significantly off-course. The violation of flight space and the rush of the military to intercept and take 'decisive action' based on incomplete data cost the lives of all 269 passengers and crew that day, in addition to the financial losses and subsequent compensation that was awarded many years later. The shooting down of this civilian aircraft by the Soviet military resulted in a homicide event.

This shooting-down of a civilian aircraft by a military power is not an isolated event. On 3 July 1988, USS Vincennes was deployed to the Arabian Gulf region. An Iran Air Airbus A300-200 passenger aircraft, flight 655, was operating a scheduled service from Tehran to Dubai Airport in the UAE, via Bandar Abbas airport in Iran. When the aircraft took-off from Bandar Abbas airport, the USS Vincennes was engaged in escort duties close to the Strait of Hormuz. The Strait of Hormuz was a location where many international ships had been attacked during the Iran-Iraq war, with numerous aircraft firing air-to-surface missiles at them. The Iran Air A300 was flying the short distance from Bandar Abbas to Dubai, descending from 14,000 ft. The electronic `friend or foe' on the USS Vincennes system identified the A300 incorrectly as an F14 Tomcat aircraft from Iran. Numerous radio messages were sent from the US warship to the aircraft using the limited equipment on board, but no reply was made. The Vincennes, already in a heightened combat state from the day's earlier engagement with the Iranian Navy, fired a surface-to-air missile, destroying the A300 aircraft and killing the 290 passengers on board. The Iranians considered the event a deliberate act of homicide, whereas the American investigation considered the events a tragic accident, noting that the warship did not have the correct communication equipment for commercial aviation. Furthermore, an international investigation concluded that while the Iran Air flight crew did hear three warnings given on distress frequencies, they did not acknowledge them, believing they were directed at another Iranian P3 Orion aircraft in the region at the time.

It is worthwhile for the reader to note that the significant difference between KE 007 and Iran Air flight 655 was that the Korean shootdown was a deliberate act to down an `intruder' aircraft leaving Soviet territory for international airspace, whereas flight 655 was a combination of computer and procedural failures, as well as misidentification and expectation failures from both the US military ships' company and the civilian flight crews.

8.3 Pacific South West Airlines Flight 1771, 7 December 1987

Pacific South West Airlines (PSA) flight 1771 was a scheduled commercial passenger flight from Los Angeles Airport (LAX) to San Francisco Airport (SFO). The background to this event was a large US operator had recently purchased and taken over the smaller Pacific South West Airlines operation. An ex-employee, David A Burke, a ticketing agent from the PSA operation, had recently been dismissed for theft of receipts. The events on Flight 1771 that transpired that day became well known in the aviation community as a hijack/murder/suicide occurrence, killing all occupants onboard in addition to destroying the aircraft. This flight was on a British Aerospace 146-200, a regional jet aircraft powered by four Textron Lycoming engines (LF-502 power plants). This aircraft can carry up to 100 passengers in various cabin configurations. The aircraft is a T-tail design, and the short single-aisle cabin results in passengers sitting in 5 or 6 abreast.

On 7 December, Burke purchased a ticket for PSA flight 1771 to travel to SFO to meet with his previous manager, under the pretence of regaining his employment with PSA. Flight 1771 was a PSA British Aerospace 146-200 aircraft: an example of this general aircraft type is illustrated in Figure 8.2.

Burke boarded this flight knowing that his ex-manager would commute from SFO to LAX and back. At the time of Burke's dismissal, Burke had failed to surrender his USAir employee credentials (e.g. ID badge, etc.). Burke used these employee credentials to bypass the normal domestic passenger security checks, as he was armed with a Magnum handgun revolver. In-flight, Burke wrote a threatening letter on an airsick bag, believed to be for his ex-manager, although it is not known if the ex-manager received it prior to his homicide.

The size of the flight deck on this type of aircraft is very small, as shown in Figure 8.3, with seating for only three occupants, namely the Captain, First Officer and an observer, where the observers' seat is immediately behind the centre pedestal, in front of the flight deck door. The passenger cabin and the cockpit are only separated by a small boarding area (indicated in Figure 8.3).

It is surmised that Burke went to the aircraft lavatory to load the handgun, as the door to the toilet was heard opening/closing on the CVR before the events unfolded. Burke shot the ex-manager twice in the passenger cabin, which was overheard by the flight crew who relayed this information to ATC. A crew member immediately went to the front of the aircraft, opened

FIGURE 8.2
British Aerospace 146/RJ100 series aircraft. (Paul Spikkers.)

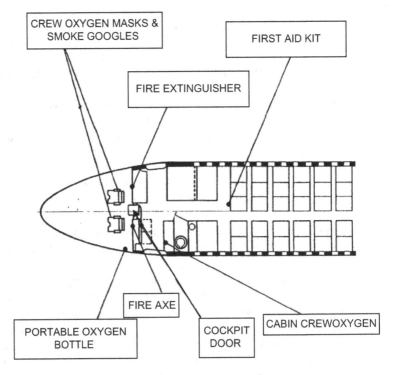

FIGURE 8.3
Drawing of 146-200 Flight Deck and forward cabin.

the flight deck door and then entered and said to the pilots, *'we have a problem.'* The Captain replied to her *'what's the problem?'* At this point Burke shot the cabin crew member dead, and said, *'I'm the problem.'* All of the audio was recorded by the CVR. Both pilots were then shot and either incapacitated or killed. The CVR indicated increasing levels of wind noise, consistent with a steep nose-down pitch attitude of around 70 degrees.

The plane descended at very high speed, at some points exceeding the speed of sound. None of the passengers were able to prevent the steep descent, and a final bullet was recorded being discharged at this time, again recorded by the CVR. The victim of this shooting is believed to be PSA's Chief Pilot, who was also a passenger on the flight. The aircraft impacted onto a hillside in the Santa Lucia Mountains at very high speed with tremendous energy. There were no survivors, and the devastation was such that 27 of the 38 passengers could not be identified due to the total carnage of the crash.

As a direct result of this hijack/homicide/suicide event, several policy changes took place. Airport security was changed to require all employees including the pilots, engineers, cabin crews, etc., to undergo the same security screening and checks as passengers. Another corporate policy change was from large organisations, to forbid travel by multiple board-level executives on the same flight, as various organisations lost multiple employees on that flight.

8.4 Silk Air 185, 19 December 1997

Silk Air flight 185 was a scheduled flight from Jakarta, Indonesia to Changi airport in Singapore. On 19 December 1997, the flight operated using a Boeing 737-300 series aircraft, carrying 97 passengers and seven crew on an 80-minute regional flight, the track being illustrated in Figure 8.4. This aircraft crashed into the Musi River in Indonesia, killing all 104 souls onboard. The NTSB investigation cites suicide by the captain, resulting in a murder-homicide event for all other occupants.

The aircraft was commanded by Captain Tsu Way Ming, a 41 year-old former Singapore military pilot, and First Officer Duncan Ward, 23-year-old New Zealander. The early phase of the flight was uneventful, and the aircraft reached the cruise altitude of 35,000 ft at around 15:53 hrs (local). At 16:05 hrs, the CVR stopped recording – which is a very unusual event as switching off this device requires an occupant to manually pull the circuit breakers, usually located in the overhead panel above the pilots. At 16:11 hrs, the DFDR then stopped recording. Again, this is a very unusual situation that requires a person to pull the circuit breaker in the overhead panel. Because there is very little CVR or DFDR recorded evidence of the final moments, it

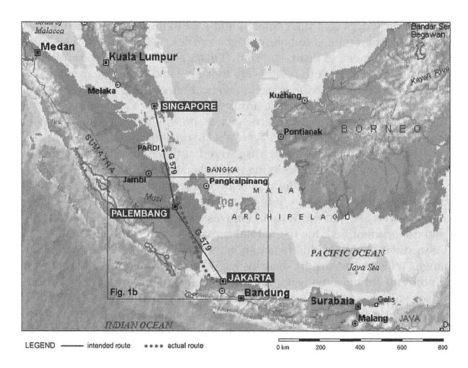

FIGURE 8.4
SilkAir's MI 185 Jakarta to Singapore flight route, showing the intended route and the actual track. (NTSC.)

is surmised that the F/O Ward was either locked out of the flight deck or incapacitated by Capt. Tsu.

At 16:12 hrs, the aircraft entered a near-vertical dive. The extreme forces acting on the aircraft from the descent caused parts of the aircraft to detach prior to impact. The aircraft hit the water at almost supersonic speeds, resulting in parts of the aircraft being embedded 15 feet below the bottom of the riverbed. The extremely high energy impact caused the whole structure, including the occupants, to disintegrate upon impact, with absolute carnage ensuing. None of the bodies of the passengers or crews were complete, and only six of the occupants were positively identified in the crash debris. The final moments of the flight and crash site are shown in Figure 8.5.

The investigation of this event was controversial. The crash investigation comprised of multiple organisations from Indonesia, USA (NTSB), Singapore and Australia. The final report was overwritten by the Indonesian Chairman, contradicting NTSB views that the evidence was consistent with a deliberate

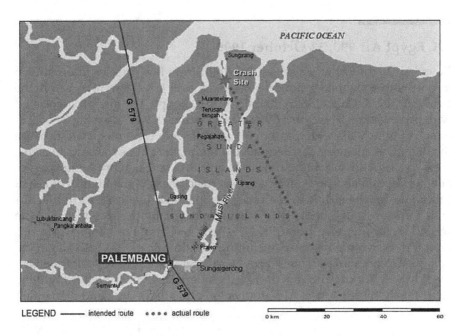

FIGURE 8.5
SilkAir's final moments and crash site on the Musi River, Indonesia. (NTSC.)

manipulation of the flight controls, most likely by the Captain. The report noted that the captain had recent financial losses exceeding $1.2 million and, in the weeks leading up to the event, the Captain had obtained a sizable life insurance policy of around $600,000. Furthermore, the day of the event coincided with a previous experience where he lost four Singapore Air Force acquaintances during his military flight training. Financial difficulty and the gain from a fraudulent life insurance claim was one possible motive for a murder/suicide homicide event for this flight.

One potential area of system failure that was identified from the accident investigation, and other historical B737 events in the USA, involved the hydraulic servo valve that is fitted to the vertical stabiliser of the B737 series aircraft. Under certain flight conditions, it was found that the aircraft's vertical surface could move to its full deflection extremity, potentially causing a rudder reversal situation. Parker Hannifin (USA), the Original Equipment Manufacturer (OEM) of these valves, lost a legal challenge raised by three families of the relatives of this SilkAir flight. A Los Angeles Court of Appeal in 2004 found that the valve manufacturer, Parker Hannifin, was liable and awarded damages of $43.6 million.

8.5 Egypt Air 990, 31 October 1999

Egypt Air flight 990 was a scheduled flight from Los Angeles International airport with a planned technical stop to refuel at JFK, New York, before the final journey to Cairo International Airport, Egypt. The aircraft associated with this event on 31 October 1999 was a Boeing 767-300 carrying 217 occupants, of which 203 were passengers and 14 were crew members. Because the flight was a long-haul scheduled operation, the airline rostered an active crew of a captain and a first officer. In addition, a relief crew comprised of another captain and first officer for the cruise phase of the flight. The usual procedure for the airline at the time was for the active crew to fly the aircraft for four to five hours before being relieved by the second crew. The second sector flight leg from JFK New York to Cairo International Airport, Egypt, is shown in Figure 8.6, including the location of the crash site in the Atlantic Ocean, some 60 miles from Nantucket Island. Shortly After departing JFK New York at 01:20 hrs (local), at around 20 minutes into the flight the relief First Officer Gameel Al-Batoui entered the flight deck and demanded

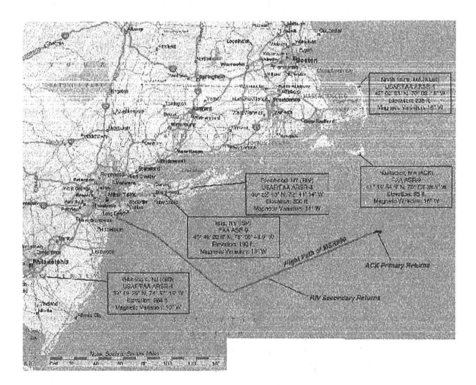

FIGURE 8.6
Egypt Air 990 second sector flight track outbound JFK to Cairo and crash site. (Egyptian CAA.)

to change the flight roster arrangements. The active First Officer initially disagreed, but later acquiesced with Al-Batoui's demands, as Al-Batoui was more senior and experienced.

The aircraft climbed to the cruise altitude of 33,000 ft, and at around 01:49 hrs it is believed that Captain El-Habashi visited the lavatory, leaving F/O Al-Batoui alone at the controls. At this point, the CVR picked out Al-Batoui saying '*I put my trust in god.*' Then autopilot was disconnected with Al-Batoui saying '*I rely on God*'. A nose-down elevator attitude was inputted by Al-Batoui, followed by the engine power levers being retarded and the engines throttled back. In this very steep descent, a zero-G nose over forward pitch was encountered, followed by an excessive descent speed, beyond the Velocity Never Exceed values. At around this point, Captain El-Habashi was able to re-enter the flight deck to try and recover the desperate situation. Electrical power was lost. Captain El-Habashi tried in vain to slow the descent by pulling back on the control column and applying engine power, asking Al-Batoui what he had done, but as the engines had been shut down, this action was not successful. The FDR indicated that the elevator circuits became disconnected at this time, because at least 50 pounds of force were applied to the separate elevator circuits. The B767 elevator control system is illustrated in Figure 8.7, showing the two separate cable circuits, the yokes

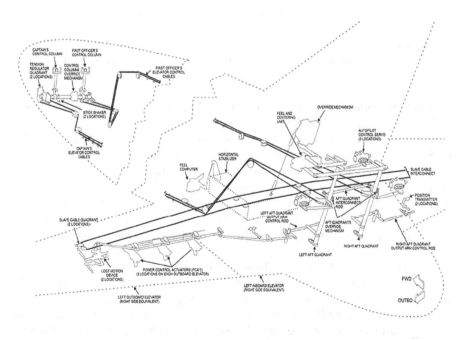

FIGURE 8.7
B767 elevator control system including components. (NTSB.)

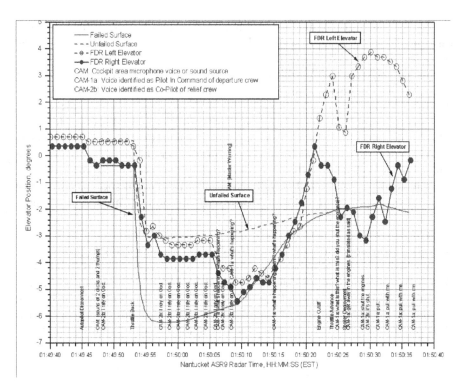

FIGURE 8.8
B767 elevator control recorded by the DFDR in the final moments of the crash. (NTSB.)

and the disconnect unit (that is included in Figure 8.7, described as the 'control columns override mechanism').

The purpose of a disconnect is to allow one elevator to function should one side of the circuit become jammed or stuck, which would give some nose attitude control in the most extreme of conditions, such as this event. However, the disconnect condition would also imply that both elevators are not being moved together in unison, such as one control being pulled while the other is being pushed. The evidence for this unusual control of the elevators in the final moments of the flight is provided from the DFDR, shown in Figure 8.8.

The event can be considered to have very strong 'indications' of a pilot-induced homicide/suicide event from the above physical evidence, although the investigation findings differed dramatically between the Egyptian civil aviation report (June 2001) and the NTSB report (March 2002). The NTSB report findings state that the fateful nose-down attitude that occurred when F/O Al-Batoui was alone in the Flight deck was not caused from a failure in the elevator control system or any other airplane failure. While investigation opinions may differ between states, the weight of evidence (from sources including the CVR and DFDR) suggests that F/O Al-Batoui deliberately caused the crash. From calmly stating the phrase '*I rely on God*' (74 seconds before the start of the decent and during the dive), to switching-off the engines and countering the Captain's desperate attempts to level the aircraft.

The F/O's actions regarding the 'probable cause' of the crash has never been established in either accident reports.

8.6 Malaysian Airlines MH370, 8 March 2013

Malaysian Airlines Flight MH370 was a scheduled night flight from Kuala Lumpur International Airport, bound for Beijing Capital International Airport in China, on 8 March 2013. The aircraft operated was a Boeing B777-200ER aircraft, carrying 12 crewmembers (pilots and cabin attendants) and 227 passengers. The aircraft involved in this occurrence was a modern B777 model, some 6 ½ years old and is still considered to have a very good safety record. MH370 went missing during its flight, and at the time of the disappearance, no wreckage was found nor any remains of the occupants.

The flight deck crew operating MH370 consisted of two crew members. The most senior pilot was Captain Zaharie Ahmad Shah, a 53-year-old pilot with 18,300 flight hours logged, in addition to being a Type Rated Instructor (TRI) and a Type Rated Examiner. In short, Captain Shah was a very experienced senior Captain with Malaysian Airlines. For this occurrence, Captain Shah is recorded as the pilot in command (P1). The First Officer was Fariq Abdul Hamid, a 27-year-old pilot with around 2,700 flight hours logged since joining the airline as a cadet second officer in 2007. F/O Hamid was still undergoing type rated flight training on this aircraft type, and the final examination was scheduled for his next commercial flight.

The aircraft took-off uneventfully at 00:42 hrs local (from a scheduled departure of 00:35 hrs), and all was normal in the early stages of the flight. ATC services instructed flight 370 at 00:46 hrs to climb from Flight Level (FL) 180 (18,000 ft) to FL 350 (35,000 ft), and at 01:01 hrs, the pilots reported to ATC that they were at this cruising altitude. At 01:06 hrs, the aircraft's ACARS data communication system transmitted the last regular data message to the ground services, as the aircraft flew over the Malaysian peninsular. At 01:08 hrs, the flight crew reported again to ATC that they were at FL 350.

At 01:19 hrs, the regional Malaysian ATC radar unit instructed MH370: '*contact Ho Chi Minh 120 decimal 9, Good night.*' MH370 made the final voice radio transmission: '*Good night, Malaysian 370.*' It is believed that Captain Shah made this final radio call, as he had done so thus far, on the approach

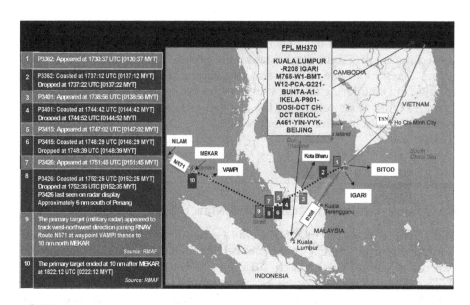

FIGURE 8.9
MH370 Flight track from Kuala Lumpur International Airport to Beijing and the point of 'loss of further communications' at waypoint IGARI. (Ministry of Transport.)

to waypoint IGARI. The scheduled flight track for the intended flight is illustrated by a solid line by Figure 8.9.

At waypoint IGARI, as the flight was leaving Malaysian controlled airspace, the procedure was for MH370 to change VHF channel (VHF 120.9 MHz) to contact Vietnams' Ho Chi Minh radar-controlled ATC service to confirm their arrival and receive instruction. The radio call to the Vietnamese ATC was never made.

At 01:21 hrs, MH370s' mode S transponder stopped transmitting data. A transponder is a device that allows pilots in-flight to input a four-digit number that helps ATC and radar better identify the aircraft. The Primary Surveillance Radar (PSR) signal is transmitted from the ground to an aircraft, and the radio waves from the ground station are reflected back to the ground radar station by the aircraft. The reflected radio wave is detected by the PSR and the aircraft is given a bearing (direction) relative to the PSR, as well as its range (calculated from the Doppler effect). The transponder equipment supplements this PSR return, which is too basic to be useful in busy skies, by reading some of the aircraft's performance data, e.g. altitude. The transponder then retransmits this attitude information with the four-digit code, which is then received by the ground receiving station's secondary surveillance radar (SSR) antenna. This additional data allows ATC to clearly identify the individual assigned aircraft by transponder code and individual flight identity. At the time of writing, the minimum current standard for transponders fitted to commercial aircraft worldwide is known as Mode S, which is fully detailed in ICAO 174-AN/110 *Secondary Surveillance Radar Mode S Advisory Circular*. Because Mode S has a digital transmission underpinning the data signal, additional aircraft performance data can be encoded and transmitted, especially with the enhanced surveillance features. In addition to the aircraft's identity, altitude reporting and flight status (e.g. on the ground or in-flight), additional data includes: the selected altitude (pilot-inputted altitude from the Flight Management Computer/Autopilot); roll angle; true track angle; ground speed; magnetic heading; indicated airspeed and Mach number; vertical rate of climb/descent; and any traffic collision service resolution advisories that have been suggested. Mode S data is particularly useful in aviation operations because of the broad performance parameters being transmitted by all operational commercial aircraft. This allows for a more automated avoidance and collision prevention service. Safety is further enhanced via software and the graphical user interfaces used by ATC services. MH370's transponder was also fitted with a functioning Automatic Dependent Surveillance-Broadcast (ADS-B) capability, which uses GPS satellite navigation data to broadcast an accurate location of the aircraft in flight to the SSR, to further enhance the situational awareness of all aircraft in-flight to avoid the possibility of mid-air collisions.

Currently, the usual procedure for commercial aviation is once the engines are started at the gate, the transponder is activated by the flight crew to start the data transmissions. The use of the transponder, being mandated, would continue until the aircraft had flown the sector, landed and taxied to the gate, whereupon the transponder would be switched-off, to the standby (STBY) position, immediately prior to engine shutdown. The transponder controls are usually located in the centre pedestal, the flat panel between the pilot's seats, as shown in Figure 8.10.

MH370 did not make radio contact with Ho Chi Minh ATC, and the timing of the switching-off (STBY) of the transponder, coupled with the time of the hand-over is very unusual, as the transponder is a required broadcast.

About the same time as the radio hand-over and switching-off (to STBY) the transponder, the Malaysian military radar detected that an aircraft in this location (IGARI) had changed direction by approximately 180°, flying back towards the Malaysian peninsular (Figure 8.11). This change of flight direction and loss of radio contact with ground-based ATC is very unusual. If MH370 were experiencing a technical problem, the normal procedure would be for the flight crew to radio ATC to make them aware of any difficulties, and then to seek the nearest airfield to land. This did not occur on that flight. Figure 8.9 shows the first major turn back towards land, which was later identified (+4 days post the disappearance) from the military radio data. During this major turn, it is proposed that the aircraft was able to cut all possible communication with the ground by the occupants of the flight deck disconnecting all the electrical power supplies. This would render all

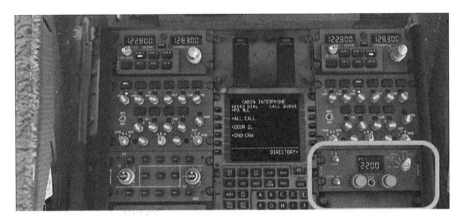

FIGURE 8.10
Image of a Boeing B777 centre pedestal showing the transponder (highlighted in grey box) transmitting 'squark' code 2200.

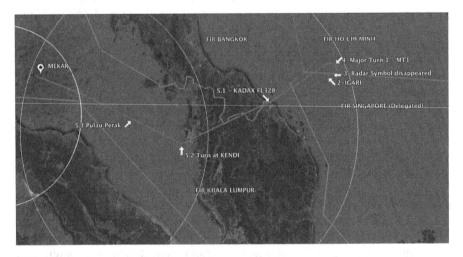

FIGURE 8.11
MH370 flight track post IGARI from military radar data. (Captio.)

electrical systems nonfunctional, including all the communication devices. The satellite communication equipment did not operate normally after 01:17 hrs local (17:07 hrs UTC). In order to disable these systems in-flight, it would have required an occupant in the flight deck to climb down a ladder into the Electronics and Equipment bay that is located below the flight deck, and to manually pull out the circuit breakers.

The task of switching-off a Boeing 777's entire electrical power system is complex, as all the back-up systems, which would usually operate automatically when the engine-driven pump generator failed, would also have to be disabled. Furthermore, to continue to fly the aircraft in this full shut down state, the pilot would have to select the emergency Ram Air Turbine (RAT) deployment, to provide emergency electrical and hydraulic power. The RAT provides limited hydraulic power to the primary flight control surfaces, enabling the aircraft to fly with limited hydraulic services - elevators, rudder, etc. Additionally, the RAT provides a limited electrical power source, but this generation capability is dependent on the load requirements and the speed of the RAT operation. Excessive electrical load drains on the aircraft's electrical system in such an emergency condition, resulting in *load shedding* - disconnecting the devices from the electrical bus bars until the RAT achieves an optimal speed.

The military radar data that was identified some four days after the initial disappearance, was published in full by the Malaysian authority's safety report, shown as Figure 8.12.

After the transponder was switched-off (STBY) at 01:19 hrs local (17:19 hrs UTC), the military radar data represented in Figure 8.11 (at 17:21 hrs UTC) shows the constant left turn to heading 273°, flying parallel to airway M765 towards Kota Bharu. The control of the aircraft, flying close to published airways, indicates that the pilot in control of the aircraft was an experienced commercial pilot, knowing that significant deviation from airways would be more easily detected by ATC controllers.

It was not until +17 minutes after (17:19 hrs UTC) the last radio transmission from MH370, that the radar controllers at Ho Chi Minh ATC decided to contact their Malaysian ATC counterpart. In the discussion at the time,

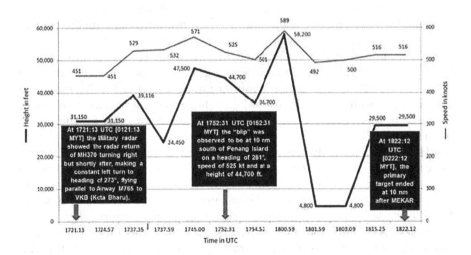

FIGURE 8.12
MH370 military radar data recorded. (Ministry of Transport.)

there was much confusion as to the whereabouts of MH370, having failed to check in with Ho Chi Minh ATC. Malaysian Airlines' Operations department provided some information regarding the estimated location based upon previous flights and tracks. Much of the confusion came from MH370 not declaring an emergency so a further four hours elapsed (c.21:00 hrs UTC) before Search And Rescue (SAR) activities could be formally raised by the relevant parties. This military radar data (Figure 8.14) indicates that after the IGORI waypoint and the 180° change of direction, MH370 then flew over the Malaysian peninsula, the aircraft climbing to an altitude of 39,100 ft. Then, MH370 descended to 24,500 ft before climbing steeply again to 47,500 ft and slowly descending to 44,700 ft at 0152 hrs local (17:52 hrs UTC).

The reader must be aware that the service ceiling of the Boeing 777-series aircraft is limited to 43,100 ft. The service ceiling is the maximum safe altitude that an aircraft can fly to when carrying passengers and loads. While it is possible to exceed the manufacturer's altitude limit, the aircraft will not perform correctly or safely. For instance, the aircraft pressurisation system and ECS (as described in Chapter 4) will not be able to function correctly. The pressure differential between the low-pressure atmosphere outside and the pressure inside the passenger cabin exceeds the aircraft's structural tolerances. The low cabin pressure will exceed the safe levels for effective respiration (breathing) as pressure will be too low for effective gas transfer in the lungs (i.e. - oxygen absorption and carbon dioxide release in the alveoli). Occupants in a low-pressure cabin environment, such as an aircraft exceeding its service ceiling, will suffer the effects of hypoxia. If not remedied with the return of pressure to safe limits, incapacitation will occur, with death following rapidly.

When the aircraft reached the coast, south of Penang (Figure 8.11, location 5.2), the F/O's mobile telephone briefly communicated with the Penang mobile telecommunication system at 01:52 hrs local (17:52 UTC). This implies that the mobile phone of F/O Hamid was switched on and communicated briefly with the ground station and nothing more. Approximately 6 nautical miles south of Penang Island, the military radar detected MH370 making a slow right-hand turn. The investigators believe that the person in charge of the aircraft knew very well the extent of the radar coverage for the region (see Figure 8.13).

MH370 continued to fly very close to published airways (Airway N571), thus the ground controllers in both Indonesia and Malaysia's Flight Information Regions would not think that the aircraft was out of place. Consequently no alarm was raised at the time. As the aircraft headed towards waypoints VAMPI and MEKAR (see Figure 8.9), the military primary surveillance radar continued to monitor the track of MH370. It was at this time, that MH370 shadowed other commercial aircraft traffic, thus masking the flight to the ATC during this passage. The operational radar limitations and ATC performance expectations appear to be fully known to the pilot in command of MH370.

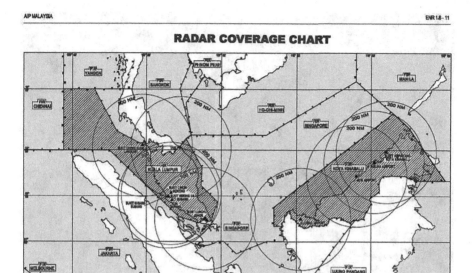

FIGURE 8.13
Accident report Radar coverage chart for Malaysia. (Ministry of Transport.)

For example, MH370 flew very close to the Malaysian and Indonesian Flight Information Region (FIR), i.e. the areas of airspace they control. Additionally, a highly experienced senior pilot would know that controllers in their different countries might anticipate their neighbour's ATC to be controlling the aircraft. The uncertainty of an aircraft flying but not showing all the data on the ATC's display would serve to reinforce this assumption.

The close proximity of Airway N571, which contained flight traffic at the time, also supports the theory that those in control of MH370 wished to 'blend in' with the other traffic. As MH370 continued to fly 'dark' and follow Airway N571, heading towards the boundaries of Kuala Lumpur's FIR and the military radar coverage, the pilot in command of MH370 appears to have turned on the aircraft's electrical systems. The purpose for turning on the electrical systems is believed to have been to make flying the aircraft `easier', to allow the use of the automated systems. As soon as the aircraft power was restored, the aircraft Satellite Data Unit (SDU) which is located to the aft of the passenger cabin, booted-up automatically. With power restored, it performed an automated `handshake/reconnection' with the Inmarsat network. This handshake is a fully automated activity, occuring at approximately 60-minute intervals. The first handshake was at 02:25 hrs local (1825 hrs UTC). The radio communication devices (including the transponders) on MH370 are believed to have remained switched-off. Further details of the Inmarsat's automated handshakes will be discussed later.

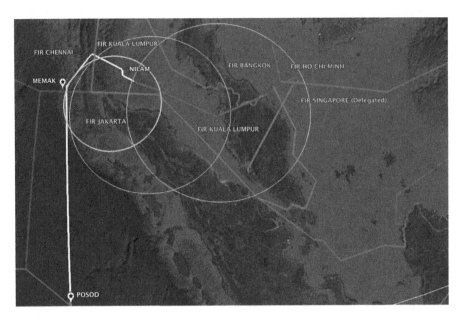

FIGURE 8.14
MH370 flight track from Palau Perak island (south of Penang) to MEMAK from military radar data. (Captio.)

Airway 571 leaves Kuala Lumpur's FIR airspace and enters India's Chennai FIR airspace, shown in Figure 8.14. It is believed that MH370 made an initial turn south west upon entering India's Chennai FIR, on the operational boundary of Indonesia's radar capability (Figure 8.14). An eye witness in another commercial aircraft was able to see the outline of a long, dark aircraft passing underneath two other commercial aircraft that were flying in the opposite direction with their navigation lights on. It is believed that the witness had observed MH370 flying underneath busy airways, again to mask the track of the missing aircraft. It is believed that just prior to the aircraft reaching waypoint MEMAK (Figure 8.14), the flight path changed again to ensure the trajectory of MH370 was heading due south, on the operational periphery of the Indonesian ground radar capabilities. While detection of MH370 by the Indonesian military radar is believed to be possible, nothing was recorded by the military.

The Inmarsat satellite handshakes became the only source of communication from MH370 for the remainder of the flight. Scientific staff at Inmarsat were able to evaluate the radio signal data received in this handshake. The scientists were able to evaluate the changes in the received satellite signals from MH370, using the Doppler effect, and combining this data with the satellites' position in orbit provided estimations of the geographical location

TABLE 8.1

Inmarsat Satellite Transmissions (Safety Investigation Report, Ministry of Transport, Malaysia, July 2018)

SATCOM Transmissions		Time	
		UTC	MYT (Local)*
1.	Aircraft departed KLIA	1642	0042
2.	Last ACARS transmission	1707	0107
3.	1st handshake – log-on initiated by the aircraft	1825	0225
4.	Unanswered ground-to-air telephone call	1839	0239
5.	2nd handshake initiated by ground station	1941	0341
6.	3rd handshake initiated by ground station	2041	0441
7.	4th handshake initiated by ground station	2141	0541
8.	5th handshake initiated by ground station	2241	0641
9.	Unanswered ground-to-air telephone call	2313	0713
10.	6th handshake initiated by ground station	0011*	0811
11.	7th handshake – log-on initiated by the aircraft	0019*	0819
12.	Aircraft did not respond to 'handshake' from Satellite Earth Ground Station	0115*	0915

* 8 March 2014

that would correspond to the MH370 transmission location. A list of MH730's Inmarsat satellite communications are provided in Table 8.1.

The estimated southward flight from the MEMAK waypoint is based on handshakes and communications (shown in Table 8.1) corresponding to predicted locations along 6 calculated arcs. Figure 8.15 illustrates the predicted arc lines of arcs two and three, with the calculated respective minima and maxima arcs, with the aircraft location falling between these two positions.

Based on the data from the satellite communications, and the additional theory from CAPITOs' report *Plausible Trajectory version 3.4 November 2019*, the predicted flight track due south is illustrated by Figure 8.16.

Based on the final arc line from the seventh handshake, the predicted search area for MH370 was determined to be in the Southern Ocean, off the western coast of Australia (Figure 8.17).

An intensive search was conducted on both the surface and the seabed once this region was identified as a potential wreckage location. The absence of a large, scattered debris field indicates that the aircraft did not exhaust its fuel supply before crashing into the ocean. Rather, no immediately visible debris indicates that the aircraft was deliberately ditched onto the sea, with the majority of the structure remaining intact. Modern aircraft contain a number of Emergency Location Transmitters (ELTs) to enable search teams to pinpoint the precise location of a crash site. Many large commercial aircraft

FIGURE 8.15
Inmarsat handshake data to correspond with seven arc lines to approximately position MH370 geographically. (Captio.)

have these transmitters pre-attached to the slide-rafts that are fitted inside the passenger doors. This Boeing 777-200 ER had eight main deck doors, with corresponding slide rafts (some fitted with ELTs). The more modern slide rafts, when they become wet with seawater, automatically activate the ELTs, transmitting a radio signal at 121.5 MHz, 243 MHz, with the newest ELTS also transmitting at 406 MHz. Furthermore, the newest ELTs are also able to transmit a pre-programmed distress message that can be detected by satellites, in addition to their precise position. As no emergency radio signal was detected at the time from the general location of the disappearance, the implication is that the ELT units were not deployed.

In July 2015, wreckage from MH370 initially was washed up onto Reunion Island. Soon after that, further debris washed ashore in Mozambique in December 2015 and February 2016. More debris was found in Mossel Bay in the Republic of South African, and Rodrigues Island in Mauritius, both in March 2016. Lastly, debris was discovered in Tanzania in June 2016. The majority of the debris found was from lightweight composite panels on the areas of the aircraft that would first hit the water in a ditching, in addition

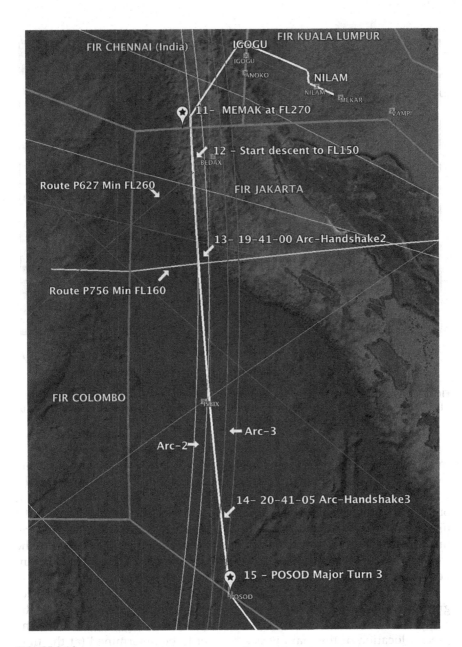

FIGURE 8.16
MH370 possible flight track including geographical arc lines 2 and 3 corresponding to the Inmarsat handshakes. (Captio.)

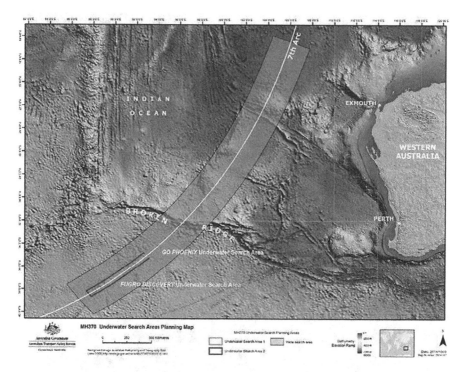

FIGURE 8.17
MH370 wreckage search area in the Southern Ocean, based on the seventh arc data provided by Inmarsat calculations. (Ministry of Transport.)

to parts from the right side wing trailing edge. If a commercial aircraft were to ditch in a 'controlled manner,' it is expected that during the impact on the water the flaps and flaperon control surfaces would detach from the aircraft on impact. Prior to ditching the aircraft, the pilot would extend the aircraft's high lift devices to fly as slowly and as low as possible to minimise the tremendous forces of landing an aircraft on water. The physical evidence of the discovered materials indicates that the aircraft ditched in a controlled manner, with the flaps extended.

A list of found parts corresponding to the aircraft location and geographical site where it was washed up is shown in Figure 8.18. While the analysis of the debris has been conducted by investigating officials, the precise location of the main wreck has yet to be determined (at the time of writing).

In the days following the initial disappearance, the Royal Malaysian Police searched the home of Captain Shah. The personal computer and flight simulation equipment (see Figure 8.19) of Captain Shah was seized by the Police from his residence on 15 March 2014.

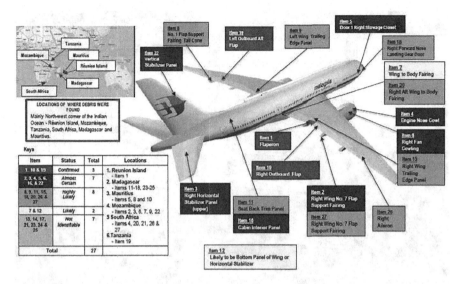

FIGURE 8.18
Location of Parts and Debris Found with respect to Aircraft. (Ministry of Transport.)

The intense international media attention regarding the disappearance of MH370, resulted in the Malaysian government making a televised press statement (18 March) regarding the ongoing investigation. Malaysia's acting Transport Minister, Mr Hishammuddin Hussein, announced that the examiners had discovered that Captain Shah had deleted all the simulation activity history in the weeks before the disappearance. Furthermore, Captain Shah was known in the home flight simulator fraternity as a keen home simulator enthusiast. The flight simulation data on the computer's hard drive was analysed. While the police forensic report concluded that *'there were no unusual activities other than game-related flight simulations'*, this finding is very limited, especially when comparing Shah's in-game waypoints to the military radar track data (Figures 8.9 and 8.11). Waypoints One to Five (Figure 8.20) bears a striking resemblance to the radar recorded flight track (shown in Figure 8.9), and these manually inputted waypoints indicate that Captain Shah was able to plan and conduct a simulated flight at his home (see Figure 8.19).

If Captain Shah had planned various scenarios using this computer, it is conceivable that he might have considered using a second hard drive to fly the simulator covertly. However, an additional hard drive in the computer was not discovered, and the hardware that was found did not indicate the removal or installation of a second drive. Nevertheless, if this was the case, Captain Shah would have understood that his simulator and hard drives would be examined carefully by the authorities after the planned event.

The police investigation also highlighted that Captain Shahs' wife and three children had moved out of the family home the day before the fateful

FIGURE 8.19
Photograph taken by Royal Malaysian Police of Captains Shah's home-based Flight Simulator. (Ministry of Transport.)

disappearance. Additionally, Shah was said to be a fanatical supporter of Malaysia's opposition leader, Anwar Ibrahim, having attended various political events, including the Ibrahim's trial and humiliation in court. Ibrahim was found guilty of homosexual offenses by the courts and was given a five-year custodial sentence that was also announced the day before the disappearance.

Very little information has been published about professional observations and concerns from Shah's airline employer. As Captain Shah had worked his entire aviation career for Malaysian Airways, it is known that Shah had very forthright views regarding politics, as highlighted above. It is perhaps unusual that the investigation reports do not cite any colleagues that had reported their concerns to this state-owned airline.

The flight and subsequent disappearance of MH370 was a complex and unclear event. For this Boeing 777 aircraft to evade detection from ATC control and military radars, significant planning and preparation would have been required. First Officer Hamid, undergoing the final stages of his type training, would not have the operational experience or the background knowledge to perform this hostile take-over with the necessary pin-point precision. As the flight showed no signs of forced entry into the flight deck and an absence of the expected distress transmission of code 7500 on the

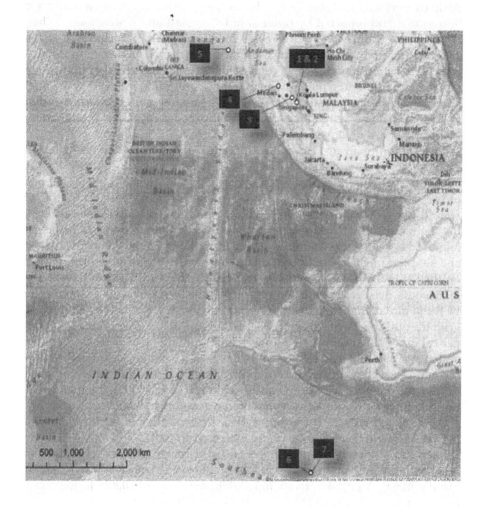

FIGURE 8.20
Captains Shah's personal computer containing Flight Simulator software with manually programmed waypoints. (Ministry of Transport.)

transponder, the focus must therefore fall on Captain Shah. Shah's considerable flying experience and detailed operational knowledge imply he knew this aircraft intimately, as his recognised company status as a TRI and TRE also demonstrates. These TRI/TRE appointments are only conferred on those pilots with the highest levels of technical understanding. Shah's home flight simulator would normally not be a cause for concern, yet it appears that Shah had planned and practiced flying the Boeing 777 in the privacy of his home. If Shah had conducted unusual flight sessions in Malaysian

Airlines' various flight simulation devices, that unusual activity would have been noted and reported by other simulation staff and instructors. Rather, Shah's home simulation device gave him an unlimited opportunity to operate the aircraft without fear of monitoring, scrutiny or detection. Shah's deliberate action to delete and corrupt the previous simulation flight data indicates an individual that wishes to remove any evidence prior to the final criminal act of hijacking the aircraft. Captain Shah would have known that F/O Hamid, on his last flight as a trainee prior to his check-out examination, would not question any order given.

For Captain Shah to take control of MH370, the most important action would be to arrange for F/O Hamid to leave the locked flight deck while in-flight. This would be easily achievable by Shah giving Hamid a menial order to visit the cabin and speak to a purser in person. As soon as F/O Hamid left the flight deck, Shah would simply need to permanently lock the flight deck door from the inside. The next action would be to incapacitate all the passengers and crews (including F/O Hamid), and the fastest and simplest method would be to manually command the air-conditionings' outflow valves to fully open (in the overhead panel – ECS control). Next, the ECS would need to be switched off, and all the bleed air valves closed (The ECS and air-conditioning operations are previously described in Chapter 4). With the outflow valves open, ECS and bleed off, all occupants in the passenger cabin would succumb to a low-pressure environment. Those who were seated and at rest had an estimated time of useful consciousness of around 30 seconds. The occupants would succumb to the immediate effects of hypoxia, despite the emergency oxygen masks deploying. At 20,000 ft, the pressure is not sufficient for respiration, even with oxygen-enriched air.

One covered switch in the flight deck overhead panel is the difference between life and death for the occupants of an aircraft.

Pilots have different oxygen systems fitted within the flight deck, and Shah would simply pull hard on the two tabs on the oxygen mask stowage, releasing the mask. A portable mask may well have been available to the cabin crew occupants (Figure 8.21) as they have a different system.

This would be a task that Shah would have performed many times in the airline's simulator training sessions, as pilots are required to demonstrate the skill of donning the mask and googles and continuing to fly the aircraft. The pilot's oxygen system is a gaseous system and can provide 100% oxygen gas with a positive pressure setting (Figure 8.22).

This means that Shah, or any other occupant in the flight deck wearing this mask, would be able to breathe normally, fly the aircraft and still function effectively in this very low pressure environment. It would also mean that any person not inside the flight deck and wearing this positive pressurised oxygen system (Fig 8.22) would be incapacitated and would expire soon thereafter. The significant change in altitude, exceeding the service ceiling after the ATC handover at IGARI, indicates Shah's determination to

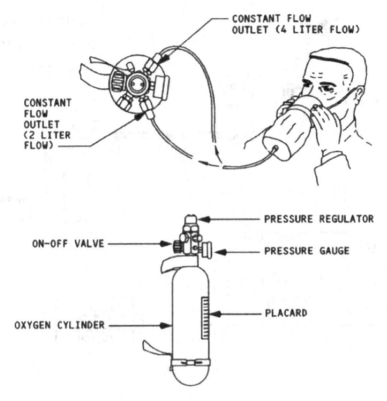

FIGURE 8.21
Portable emergency oxygen mask and gaseous bottle. (AAIASB.)

eliminate all other occupants, as some crew members in the cabin may have tried to use portable gaseous oxygen breathing systems.

The full facts pertaining to the loss of MH370 and the potential homicide of the passengers and crew will probably never be established. In time, the wreckage will probably be identified on the seabed. Likewise, with the location of the sunken structure, the black boxes (DFDR and DCVR) will be removed and analysed. However, if MH370 was able to be deliberately flown without power, it is equally feasible that Shah could have wiped the data from these devices by replicating a Weight On Wheels scenario in conjunction with the park brake set to ON. Additionally, it would have been possible to reach up to the overhead panel and isolate the electrical power circuit breakers for both DFDR and DCVR, thus removing all traces of the evidence.

Unfortunately for Malaysian Airlines, the disappearance of MH370 would not be the only tragedy to affect the carrier that year. Later that year, on 17 July 2014, MH17 would be lost. MH17 was a scheduled passenger flight from

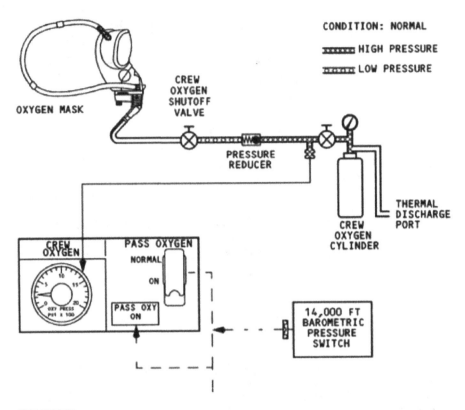

FIGURE 8.22
Flight deck emergency oxygen mask including smoke goggles and microphone. (AAIASB.)

Amsterdam International Airport in Holland to Kuala Lumpur, Malaysia. The operational aircraft was a Boeing B777-200ER, carrying 283 passengers and 15 crew members. The operation of MH17 was routine, and no fault or criticism can be directed to the crews or passengers in this aircraft loss, as the aircraft was deliberately shot down using a surface to air missile system while overflying Hrabove, Ukraine, close to the Russian boarder. At the time of the shootdown, Ukraine was undergoing a civil war with Russian-supported separatists. The Russian military supplied the separatists with a modern 'Buk' surface-to-air missile defence system that originated from the Russian military. MH17 was cruising at 33,000 ft, very close to airway L980, under the control of the Ukrainian ATC, when at 13:20 hrs UTC, a missile reached the aircraft and exploded, killing all the souls on board and downing the airline in the disputed and hostile Ukrainian territory. Immediately after the shootdown, the pro-Russian separatists returned the Buk missile system back to the Russian military, in an attempt to cover their tracks. The Russian military consistently denied all involvement in this incident, including the loan of missile equipment. The subsequent air-crash investigations (as per ICAO requirements) and the prosecution based investigation were both lead by the respective Dutch Authorities. This established that this was a deliberate act of homicide on the part of the pro-separatists, but those involved have yet to be brought to trial due to complexities involving the state of Russia hampering this action. Precisely why MH17 was shot down remains unanswered.

8.7 LAM Mozambique Flight 470, 29 November 2013

Linhas Aéreas de Moçambique, or LAM Mozambique, operated a scheduled flight from Maputo International Airport, Mozambique, to Quatro de Fevereiro International Airport, Angola. On 29 November 2013, the scheduled flight of an Embraer 190 aircraft was flying this sector, with two pilots, four cabin crew and 27 passengers. The Captain (P1) at the time was Herminio dos Santos Fernandes, a 49-year-old experienced pilot with around 9,000 flight hours logged. The First Officer (P2) was Grácio Chimuquile, a much younger 24-year-old pilot with 1,400 flight hours.

While the aircraft was en-route to Angola, flying at an altitude of 38,000 ft over Botswanan airspace (Figure 8.23), the aircraft entered a steep dive and crashed, killing the occupants. Although the descent was started in Botswanan airspace, the impact took place just inside the Namibian border in the Bwabwata National Park.

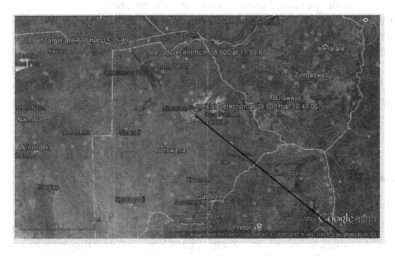

FIGURE 8.23
Flight track of LAM 470, including start of decent. (Ministry of Works and Transport.)

The event is formally recorded in the Namibia Directorate of Aircraft Accident Investigation report (part of the Ministry of Works and Transport) as a pilot homicide/suicide event.

The early phase of the flight was normal, but at 1 hour 50 minutes into the flight, the CVR records F/O telling the Captain that he requires the bathroom, and if the Captain can take over the controls (P1). The Captain replies, *'No problem'*. The CVR records various sounds in the flight deck including the door being unlocked, and immediately locked once the door closed. About three minutes later, the CVR records sounds of clicking, which the crash investigation report attributes to being the altitude autopilot pre-selector being changed. This pre-selection was a deliberate action that commanded the deep dive. Two minutes later, the CVR records sounds of thumping, believed to be from the cabin, probably the F/O desperately trying to re-enter the flight deck unsuccessfully.

The aircraft crashed inside the Namibian Bwabwata National Park at very high velocity, exploding upon impact. The CVR and DFDR recorders were both recovered in the search and rescue phase, although due to the nature of the event, there were no survivors of this pilot homicide.

The suicidal actions of the Captain have been considered by many investigators, but all the reports include the Captain's personal problems in the time running up to the event. These problems included the Captain's son's suspected suicide in November 2012, his daughter undergoing heart surgery at the time of the crash and his ongoing divorce proceedings. Furthermore, the Namibia Directorate of the Aircraft Accident Investigation report cites that the Captains' 'financial particulars' were not provided by the Mozambique authorities. The assumption that can be inferred from this omission is that the suicidal Captains' personal were also a contributing factor.

In summary, the event demonstrates the strength of the post-9/11 reinforced flight deck door, and once the door is locked from the inside, there is very little that can be done from the passenger cabin to regain access to the flight deck. While the introduction of Sky Marshals and secured flight decks appears to have stopped passengers attempting a terrorist take over, the risk of multiple homicides (instigated by one of the flight crew members) has been increased.

8.8 Germanwings 9525, 24 March 2015

Germanwings 9525 is perhaps the most famous and widely publicised pilot homicide/suicide event in recent years, caused by the psychotic F/O. Germanwings Flight 9525 was a return flight from Barcelona-El Prat Airport, Spain, to Dusseldorf Airport, Germany. The aircraft flown was an A320-200 series, carrying 144 passengers and 6 crew (2 pilots and 4 cabin crew). The pilot in command was Captain Patrick Sondenheimer, a 34-year-old pilot with around 6,000 flying hours, and about 3,800 in this type of aircraft. The F/O was Andreas Gunter Lubitz, a 27-year-old very junior pilot with a total of 919 flight hours, 500 on the Airbus type, who had a significant history of clinical mental health illnesses. During the return flight from Spain to Germany, as the aircraft overflew the French Alps, F/O Lubitz took over the aircraft by locking Captain Sondenheimer out of the flight deck, and deliberately crashed the aircraft into the mountains. There were no survivors, and the high-speed impact into the mountains completely destroyed the aircraft. The subsequent accident investigation was led by the French Bureau d'Enquetes et d'Analyses (BEA) accident investigation authorities, with their final report published in March 2016.

Lubitz started his flight training at Lufthansa's flight training school in Bremen, Germany on 1 September 2008. On 5 November 2008, Lubitz's training was suspended for medical reasons – and was hospitalised for a severe depressive episode. Lubitz underwent treatment for his mental health, but on 9 April 2009, his class 1 medical license was not revalidated by the Lufthansa Aeromedical centre '*due to the depression and the medication* (Lubitz) *was taking for it*'. Lubitz appealed the decision, obtaining a new class 1 medical certificate on 28 July 2009 (valid until 9 April 2010) with the endorsement: '*Note the special conditions/restriction of the waiver FRA 091/09-REV*'- implying that because of this previous psychotic episode, any reoccurrence would render the medical license to become invalid. Lubitz restarted his flight training in late August 2009, including further flight training in Phoenix, Arizona, USA from November 2010 to March 2011. From September to December 2013, Lubitz attended and passed his A320 type rating at Lufthansa's centre in Munich, Germany, and joined Germanwings on 4 December 2013. Further company flight training was undertaken by Lubitz from January to June 2014, and on 26 June 2014, he was appointed as a co-pilot. Lubitz' training and proficiency checks showed that the instructor/examiners considered him to be above standard, and no records or observations were made by staff or colleagues to indicate any concern about his attitude or behaviour.

Because of Lubitz' complex medical history, he found it very difficult to obtain loss of license insurance, which was contracted to all Lufthansa and Germanwings pilots should they become unfit and unable to continue to fly. This insurance was deemed necessary, because Lubitz had needed to pay around €60,000 to undergo the flight training and had a loan of €41,000 outstanding. Lubitz wrote (December 2014) in an email that his medical background was hindering his ability to obtain this required insurance policy. In December 2014, Lubitz appeared to relapse, visiting several doctors and a psychiatrist on at least two separate occasions. He was prescribed antidepressant medication, some 5 months after his last class 1 medical revalidation certificate. Later, in February 2015, Lubitz visited a private physician who diagnosed a psychosomatic disorder in addition to an anxiety disorder, referring Lubitz to a psychotherapist and a psychiatrist. The medical conditions persisted, so on 10 March 2015 Lubitz visited the same private physician again, except this time the medical physician-diagnosed possible psychosis and recommended psychiatric treatment. It was later discovered that multiple sick leave certificates were issued to Lubitz, but none of these were forwarded to Germanwings.

Some 14 days after the final physician's visit, the events surrounding Lubitz and his medical history would show the weakness in the airline safety and security system, resulting in a wilful homicidal and suicide act by F/O Lubitz during Germanwings Flight 9525. In the subsequent investigations, it was discovered that Lubitz had visited over 40 different doctors in the previous five years, complaining of various medical problems. Lubitz was convinced that his license would be revoked once his full medical history and condition was established, thus losing the ability to work and fly. At the time of the event, the BEA final report highlighted that the German regulations for personal privacy regarding medical history was unclear with regard to *when a threat to public safety outweighs the requirements of medical confidentiality*. The lack of communication between the medical providers and the medical licensing authorities, specifically the LBA, was incomplete at the time of the event, and incompatible with European requirements.

The outbound sector from Dusseldorf to Barcelona was uneventful. The return sector took-off from Barcelona at 10:01 hrs local (09:01 hrs UTC), with the complement of 144 passengers and 6 crew, scheduled to land at Dusseldorf at 11:39 hrs local (10:39 hrs UTC). The pilot flying this sector was F/O Lubitz, with the Captain undertaking pilot-not-flying (P2) duties. The aircraft reached the top of cruise at 10:27 hrs local (09:27 hrs UTC), around halfway between the Spanish and French coast over the Mediterranean sea. The last radio communication was the Captains' read back acknowledgement to ATC at 10:30 hrs local (09:27 hrs UTC), and at this point the Captain asked F/O Lubitz to take-over the radio because he needed to leave the flight deck to visit the bathroom. The DCVR captures at 10:30 hrs 24 seconds local (09:30 hrs 25 s UTC) the sound of the flight deck door opening, and three seconds later closing. Figure 8.24 illustrates the A320 reinforced flight deck door structure, in particular showing the multiple locking mechanisms.

No sooner had the Captain left the flight deck, it is believed that F/O Lubitz immediately deadlocked the flight deck door from the inside (Figure 8.25), by moving the toggle switch on the centre pedestal from the NORM (normal operation) position, to the LOCK (deadlocked) position.

By moving this switch, entry from the passenger compartment back to the flight deck using the keypad sequence was impossible; Lubitz was secure and alone in the flight deck. At 10:30 hrs 53 seconds local (09:30 hrs 53 s UTC) F/O Lubitz changed the autopilot input from 38,000 ft to the minimum value of 100 ft (using the Flight Control Unit – Figure 8.26) in one second.

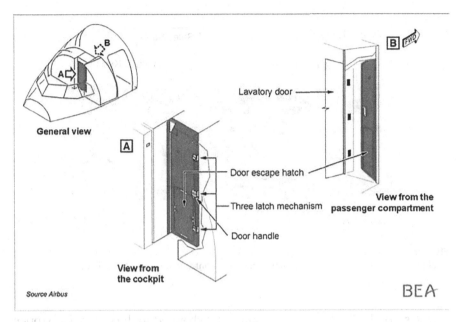

FIGURE 8.24
Airbus A320 reinforced flight deck door. (BEA.)

FIGURE 8.25
Airbus cockpit door locking controls located on the rear left of the centre pedestal. (BEA.)

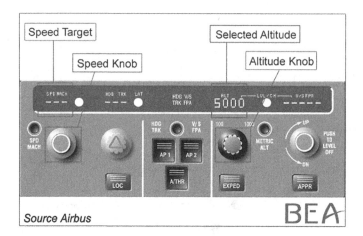

FIGURE 8.26
Flight Control Unit located on the Glareshield that controls the autopilot. (BEA.)

One second later, the aircraft's autopilot changed to the OPEN DES (open descent mode) and the autothrust was changed to THR IDLE mode. The aircraft then began the descent, and the power to both engines was decreased. At 10:33 hrs local, the aircraft was descending between 1,700 ft per minute to 5,000 ft per minute. During this rapid and unexpected descent, ATC called Flight 9525 to ask what cruise level they had asked for, but there was no response. Around this time, the aircraft was passing 30,000 ft and the ATC controller again attempted to make radio contact twice more unsuccessfully. At 10:34 hrs local (09:34 hrs UTC) the buzzer request to enter the flight deck was made, followed by knocking on the door (recorded on the DCVR). Numerous other attempts from inside the cabin were made to gain entry to the flight deck, all unsuccessful. ATC continued in vain to attempt to establish communication, yet no reply was made. F/O Lubitz' breathing was recorded during this descent – nothing was spoken further to the Captain leaving the flight deck. At 10:41 hrs 06 seconds local (09:41 hrs 06 s UTC) the aircraft impacted the mountains at Prads-Haute-Bleone in the French Alps at high speed (Figure 8.27).

There were no survivors, and although the wreckage was quickly found, the largest piece of the wreckage was 3 to 4 meters long, indicating the ferocity of the impact.

During the subsequent accident and criminal investigation, it was discovered that F/O Lubitz had used his personal tablet computer to conduct web-based searches including threats such as *ways to commit suicide* and *cockpit doors and their security provisions.*

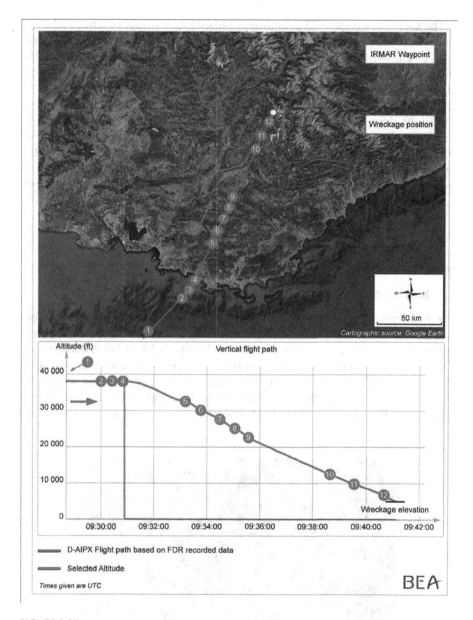

FIGURE 8.27
Germanwings 9525 accident trajectory, including Point 3, where the Captain leaves the flight
deck prior to the deliberate crash. (BEA.)

The Germanwings 9525 pilot homicide event changed commercial aviation significantly. All national regulators (national aviation authorities) and employers (airlines) re-evaluated the psychological profiles of the pilots, looking for similar tell-tale signs in other flight crews that held class 1 or 2 medical certificates. Rules and procedures were immediately changed to ensure that in the event that one of the pilots wishes to leave the flight deck to visit the toilet, a cabin crew member will be called to the flight deck in advance of this absence, to potentially prevent another possible pilot homicide occurrence. However, Flight 9525 illustrates the lengths a determined individual pilot will go to hide their mental health problems, including contemplating multiple homicides and their own suicide.

8.9 Horizon Air (theft/suicide), 10 August 2018

On 10 August 2018, at Seattle-Tacoma International Airport, Washington, USA, Richard Russell, a 28-year-old airline ground service agent used his credentials to gain access to a parked DeHavilland DCH Dash 400 aircraft. Russell made an unauthorised take-off (considered to be theft), flew around the locality for around 1 hour 15 minutes, and later committed suicide by intentionally crashing the aircraft into Ketron Island in the Pugent Sound region (see Figure 8.28). This event illustrates the ease at which employees

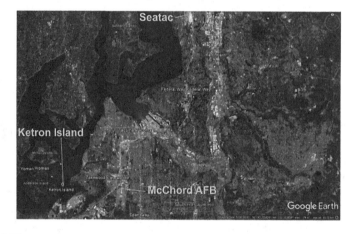

FIGURE 8.28
Theft of DHC Dash 8 Q400 from Sea-Tac Airport, and subsequent deliberate suicide on Ketron Island. (Google Maps.)

of airlines can move in restricted areas with little or no supervision, gain access to large commercial aircraft, and steal them with little or no formal pilot training.

Richard Russell was an experienced Horizon Air ground service agent, being employed as part of the ground staff 'tow team', i.e. towing and moving aircraft on the ramp and airport apron.

Because this event was a deliberate act of theft (with the potential for other serious activities, such as terrorism), the subsequent investigation was conducted by the Federal Bureau of Investigation (FBI), supported by the NTSB.

The FBI concluded its report on 9 November 2018 reporting the following findings.

On 10 August, Russell arrives at 14:36 hrs (local) at the Port of Seattle security checkpoint, in accordance with his scheduled work shift. At 14:38 hrs, Russell clears the security checkpoint and goes about his duties. No anomalies are noted at this point.

However, at 19:15 hrs, Russell arrives in a tow vehicle at the Cargo 1 location, at the far end of the Sea-Tac airfield. At 19:19 hrs Russell enters Horizon Air DHC dash 8 -Q400, registration N449QX. This appeared normal thus far. However, at 19:22 hrs Russell begins the engine start sequence on this aircraft – an activity that a ground service agent would not be authorised to undertake. The engines start successfully, and the propellers start to rotate. At 19:27 hrs, Russell climbs out of the 'live' aircraft with the engines still running. Using the ground tow vehicle, he positions the aircraft with the nose pointing towards the airfield. Russell re-enters the aircraft at 19:28 hrs, and at 19:32 hrs the aircraft starts to taxi, away from the parking location. At 19:33 hrs, Russell takes-off in the aircraft.

The FBI concludes the report by stating that at 20:46 hrs, the *'DFDR shows the end of flight, known to investigators as the aircraft crash on Ketron Island in Pierce County, Washington'*. The general area is represented in Figure 8.28.

In the moments immediately after the unauthorised take-off, Seattle-Tacoma ATC attempted to make radio contact with the flight and occupant. Initially, there was no response from Russell, but later Russell engaged in a long conversation with ATC. Russell asked questions including the suggested altitude to commence a barrel roll in the stolen Q400. After Russell had completed these unusual aerobatic manoeuvres, ATC attempted to convince Russell to attempt to land the aircraft at a nearby runway, to which Russell declined. Russell realised the magnitude of his actions, and his conversation included reference to this. Some 45 minutes into this unauthorised flight, the military scrambled two armed interceptor F-15C Eagle aircraft from Oregon's 142nd Fighter Wing. The F-15s are filmed by local residents in the area, closely following this stolen Q400 aircraft. The F-15s initially attempted to direct the aircraft out towards the Pacific, but Russell did not follow this instruction: the interceptors did not shoot down the aircraft.

In the final minutes of the flight, when ATC pressed Russell again to land the aircraft, Russell replied, '...*wasn't really planning on landing it*'. This confirms that Russell had the intent of crashing the aircraft, with very high probability of suicide, thus avoiding the legal consequences of his actions that day.

At 20:46 hrs, Russell deliberately crashed the DHC-Dash 8-Q400 aircraft into Ketron Island, an act of suicide that destroyed the aircraft. No other persons were injured.

This event demonstrates the ease by which ground staff can move around the apron, often unsupervised, to enter aircraft, move aircraft and even gain access to flight decks to follow aircraft checklists and subsequently start the engines. While some commercial aircraft are fitted with external key-based locks to the passenger and cargo/engineering compartment doors, these locks are very rarely used. For aircraft with a maximum take-off mass greater than 5,700 kg, there are no key switches required to switch on the electrical power from the batteries; no key-start for the Auxiliary Power Unit (APU) or key-starts for each of the respective engines.

In summary, any person such as an employee, passenger or intruder on the airport's apron or ramp area can gain access to any large commercial aircraft. They can enter any aircraft via the external steps or an airbridge, proceed to the flight deck, open the door, follow a laminated checklist and start the aircraft's engine. Lastly, they can potentially make an unauthorised take-off as demonstrated by the HorizonAir theft.

Another less well-publicised event occurred in 2000 at London Gatwick (UK) Airport during a weekend. A short background: all airports around the world have wildlife problems, and Gatwick Airport is no different. For example at Gatwick, rabbits breed at an incredible rate, and foxes enter the airfield perimeter to hunt and consume the rabbits. The grass that surrounds the airfield's taxiways, active runway and emergency runway all have 'personnel detectors' sited in the grass. However, with the rabbits and foxes demonstrating the 'circle of life' during the hours of darkness, the airport had become accustomed to the spurious detection systems activating. Other wildlife, including badgers, were also known to activate the detection systems in the grass on a regular basis.

Two local men, while heavily under the influence of alcohol, climbed a security gate, close to Hanger 6 and Hanger 3, at around 3 am on the southern side of the airfield (see Figure 8.29, thick black line represents the intruder's route from the southern perimeter, over the runways to the cargo and maintenance areas).

Having climbed over the fence, they walked the short distance of around 100 m, across the grass, crossed the active runway, the emergency runway and the various taxiways to the cargo terminal location. At the cargo terminal (Figure 8.29, top left of image), they saw a parked aircraft that had metal ground steps pushed up to the aircraft door. They opened the aircraft door

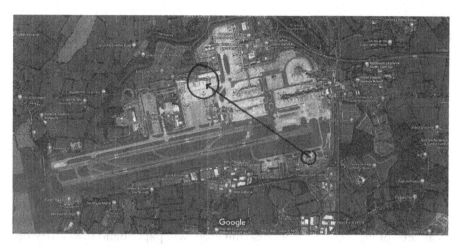

FIGURE 8.29
Plan of London Gatwick airport showing the path of the drunk intruders. (Google Maps.)

(as the plane was unlocked) and then proceeded to the flight deck. Although the *personnel detector*s in the grass had been activated, the ground staff had dismissed the activation as another wildlife event. It was only when a maintenance ground staff member reported to security that unknown persons were seen around the cargo terminal area, that an urgent security search was initiated. The search found the two drunk persons in the flight deck of the parked aircraft, having been operating the switches in the flight deck *to see what would happen*. Both individuals were arrested and charged with the criminal offence of entering an airport. Fortunately, on this occasion, the perpetrators did not have hostile intensions of bringing harm to the aircraft, themselves or others.

8.10 Other Current Sources of Weaknesses in Commercial Aircraft

This chapter thus far has detailed how some protective measures have been introduced further to pilot homicide events. For example, after the 9/11 attacks, the flight deck doors have been strengthened with ballistic-resistant panels, electronic locks, and an ability to deadlock the door from the inside. However, there are some current areas of weakness, affecting the safety of the occupants that remain unaddressed.

8.10.1 Aircraft Toilet Smoke Detectors

Large commercial aircraft are fitted with aircraft toilets. Each toilet is fitted with a smoke detector in the ceiling compartment, and above the waste bin there is a halon-type fire suppression system. The fire detector typically includes an ionising detection cell inside the unit to provide a positive detection for smoke being present. If a positive signal (i.e. smoke) is detected, the unit sounds an audible alarm. The aircraft's communication system will call the cabin crews (via a chiming signal in the galley/crew stations). Outside the toilet a light on the bulkhead will illuminate, and lastly a signal will be sent to the flight deck. The system and general bathroom layout are illustrated in Figure 8.30.

Because the aircraft operates in a very low-humidity environment, all paper products become dry and flammable. The waste-paper bin in the toilet is an area of great concern, because the used, wet paper hand towels dry out during the flight, and if a passenger attempts to smoke a cigarette in the toilet compartment, should a match or hot cigarette butt be placed in the bin, the paper towels would very easily combust. Under the toilet's countertop, is a waste-paper bin. Directly above this bin storage is a halon extinguisher. The halon unit is about the size of a tennis ball, with two metallic (sealed) pipes pointing downwards about the bin. If a fire takes hold in the bin, the material in the pipes that seals them begins to melt at around 170°F/77°C, and the halon will automatically discharge into the bin, extinguishing the fire. A visual gauge may be present on the halon extinguisher bottle, likewise, a temperature-sensitive strip may be included on the outside of the unit, to visually indicate to the crews that the temperate under the counter (above the bin) has exceeded given values.

The procedure for a toilet smoke detection event is for cabin crew to immediately identify the affected toilet compartment (as a priority action), bring a BCF extinguisher with them and check the toilet door surface (with the back of their hand) for heat. If the door is hot, they will open the door (from the outside using the unlock mechanism), crack the door open and fully discharge one whole BCF bottle into the toilet, closing the door afterwards. At the same time, the flight crew will usually initiate a descent and make an unscheduled landing, due to the cabin fire situation.

While smoking in the toilets is a prohibited event – often with criminal sanctions upon detections - this does not stop some passengers from doing so. Passengers have interfered with the ceiling smoke detectors over the past 30+ years (from the start of the in-flight smoking ban). Methods to obstruct the ceiling smoke detector have included: covering/wrapping the detector in clingfilm; using wetted toilet tissues/wetted hand towels/spraying the detector with a layer of shaving cream; physically damaging the ceiling detector with the aim to render it inoperable, etc. Some of these methods

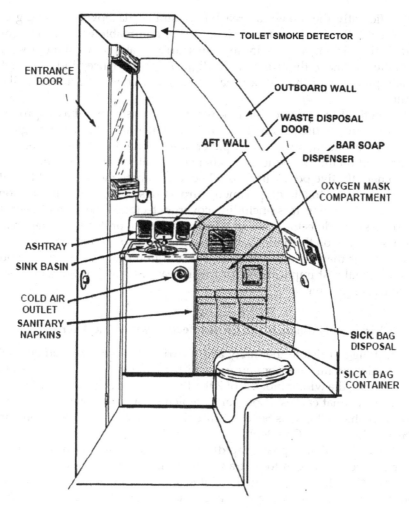

TOILET SMOKE DETECTOR

ENTRANCE DOOR

OUTBOARD WALL

WASTE DISPOSAL DOOR

AFT WALL

BAR SOAP DISPENSER

OXYGEN MASK COMPARTMENT

ASHTRAY

SINK BASIN

COLD AIR OUTLET

SANITARY NAPKINS

SICK BAG DISPOSAL

SICK BAG CONTAINER

FIGURE 8.30
Aircraft toilet including smoke detection and wastebasket fire suppression systems. (NTSB.)

have been successful in causing the ceiling detector unit to be unable to detect smoke.

The weakness is as follows: if a determined individual wishes to commit mass homicide and suicide in-flight, said perpetrator could disable or interfere with the ceiling detector, deliberately starting a fire within the toilet compartment. As previously mentioned, toilets contain quantities of paper toilet rolls and paper hand towels. If the fire were able to become hot enough, the non-metallic (plastic) interior materials in the compartment would combust. If the temperature of the fire in this enclosed space were to rise sufficiently, the fire would reach the point of 'flashover', meaning that all adjacent flammable materials would emit flammable vapours and combust. Extinguishing such a fire in a pressurised cabin environment would be highly unlikely, the previous in-flight fire experiences indicate that a catastrophic loss of the aircraft, passengers and crew would be the likely result.

One method to mitigate this risk would be to use paper-based products that are manufactured with fire retardant materials. This technology has been used in the manufacture of paper-based confetti for many years, and the application this 'knowhow' to other absorbent papers is possible.

Another effective possibility would be to eliminate the combustible materials used in the toilet compartment furnishings, and to install additional smoke and heat detectors (including covert detectors). The current overreliance on a single detector is not an effective means of protection: aviation as a business is built on the philosophy of having layers of protection and redundancy, rather than a single solution. The addition of more detectors would have minimal cost implications for the increase in mass, or the total cost of the toilet compartment.

8.10.2 Viral Contamination and Cross Infection within the Aircraft

The next biggest risk to the safety and security of passengers and crew in a commercial aircraft are from airborne viruses and potential cross-infection risks. During the winter months of 2019, the emergence of a new, highly contagious virus that causes respiratory problems was investigated and identified (in Wuhan, China) as Severe Acute Respiratory Syndrome Coronavirus 2 (SARS-CoV-2) or 'COVID-19'.

COVID 19 initially spread rapidly in the Wuhan region of China, and latterly infections began to be transmitted through the carriage of infected persons. With the modern, highly interconnected mobility of the world's population, rapid long-distance travel (via airlines) has allowed the infection to be carried across the globe. Infected members of the public may not be aware that they have COVID-19, and thus the spread of this virus across the world was categorised as a pandemic, reluctantly, by the World Health Organization on 11 March 2020.

Cruise ships use a similar Environmental Control System to that of commercial aircraft, with the exception that there is no difference in air pressure between the inside and outside of the vessel. The air provided to the ship is conditioned, mixing quantities of fresh air with recycled ventilation air.

COVID-19 began to spread outside of China in early 2020, where an elderly Chinese passenger (with a cough that developed the day before) flew from Hong Kong to Tokyo on 17 January 2020 for a Lunar New Year holiday aboard the Diamond Princess cruise ship. The passenger departed the cruise on 25 January, when the Diamond Princess docked in Hong Kong. The cruise ship then sailed on to the Yokohama port area, and on 1 February 2020, the Japanese port authority placed the ship and all the passengers aboard in a state of quarantine, once the Chinese passenger that departed some days earlier tested positive to this new COVID-19 virus. Over the next month, the Diamond Princess ship, still holding 3,711 quarantined passengers, saw the virus spread through the ship's community to 712 persons, with at least one recorded death.

A similar aviation passenger contamination event occurred on 25 August 2020, on TUI Flight 6215 from Zante, Greece to Cardiff Airport, UK. The flight appeared normal for the 193 passengers, yet a week after landing it emerged that seven separate cases of COVID-19 had been detected. All the passengers were ordered to isolate pending further testing, and it was later discovered that the infection had spread to sixteen individuals, located throughout the aircraft. The close proximity of passengers in the aircraft, coupled with the ventilation system, allowed the infectious disease to spread to other travellers.

The European Union Aviation Safety Agency (EASA) has issued various guidance materials on the aviation health safety protocols in relation to the COVID-19 pandemic. The management of passengers onboard the aircraft highlights that the *'guidance materials previously implemented for the Middle East Respiratory Syndrome Coronavirus can be used as a baseline, as the scientific evidence on COVID-19 in-flight transmission is still lacking'*. EASA has recommended that aircraft limit the use of the recirculated (recycled) air flows, and where possible, run the environmental control system in the maximum flow settings, thus providing maximum flows of fresh, clean conditioned air into the passenger cabin.

Note: It is also important to note that viruses have a natural tendency to mutate: COVID-19 in its current form will cause significant further disruption from the initial outbreak in December 2019 to well beyond Autumn 2021. The virus is known to mutate and change, therefore (at the time of writing) the future levels of infection, contagion and mortality will not be fully understood further testing and research has been undertaken. That said, the next variation of COVID-19, based on the current scientific observations regarding strains and mutations, will have an equally devastating effect on the transportation industry.

In light of the rapid cross-contamination experienced by passengers on cruise ships that use similar environmental control systems (ventilation), it is not beyond the realm of possibility for a similarly infectious disease to be deliberately carried onboard by an infected 'passenger'. Deliberate infection by the traveller, knowing the transmission rate and subsequent effects, would have a maximum effect not on the flight, but rather on the occupants in the days and weeks that followed. The potential psychological effects of such an event could be just as devastating to the broader public as a total loss of the aircraft, perhaps more so, because the lack of clarity and absolution would affect a much larger number of persons.

In mitigating the airborne contamination of the aircraft environmental control system, it may not be possible to completely avoid the use of recycled air sources. To minimise the effects of the current COVID-19 (or similar) being recirculated continuously through the ventilation system, the aircraft manufacturer's will need to modify their ventilation systems, to include different 'clean room' filtration technologies. The exclusive and overreliant use of High-Efficiency Particulate Air (HEPA) filters is not a wise sole solution. If hospitals with infectious wards and virus testing laboratories use a combination of filtering technologies to remove airborne pathogens, then the ventilation manufacturers should also mirror this strategy and consider the inclusion of high energy photonics, namely germicidal ultraviolet, in addition to electrostatic filters, etc.

8.11 Conclusions

This chapter has highlighted a number of key homicide type events from the 1980s to present day. A number of military forces have coordinated and carried out deliberate shootdowns of commercial aircraft, some due to the positioning of the defending state, others due to computer misinterpretation of the incoming event. Both scenarios have resulted in the deaths of all the crews and passengers and the immediate loss of the aircraft.

For multiple homicidal acts by pilots against their passengers, a common theme has been the financial difficulties of the perpetrator, with the potential gain coming from the perpetrator being able to abscond from their financial liabilities.

The mysterious disappearance of MH370 still remains one of aviation's greatest unanswered questions. To covertly take an aircraft by force requires planning, deep understanding and aptitude. The financial implications are present, but a deep and subversive political motivation may well have been the tipping point. Likewise, Germanwings 9525 illustrates the lengths some individuals will go to conceal serious underlying mental illnesses, possible financial woes and their ambition to commit suicide, dispatching their fellow passengers and crew via their final act.

Lastly, the theft of aircraft and subsequent suicides have shown to the world the security implications of ground employees allowed roam unchecked inside an airport. Furthermore, staff and intruders can enter aircraft with ease, gain access to the flight deck, and using the available checklists, start the engines to steal the aircraft.

The next chapter will present how current commercial off-the-shelf technologies can be applied to modern commercial aircraft, to prevent deliberate pilot homicide acts. The solution for an aircraft under siege from a mentally disturbed pilot cannot be scrambling interceptor aircraft with air-to-air missiles, because the potential perpetrator is already prepared to die.

9

Minimizing Loss: Modifying Current Aircraft and Processes

9.1 Introduction

This chapter will explain that ground-controlled aircraft, as a technology, have roots that go back almost to the start of aviation itself. While the modern media might attribute the application of drones and ground-controlled reconnaissance aircraft to the period after the 1990 Gulf War, the technology behind this has been secretly developed by the military going back to the Great War (1918). Some full-scale testing of a ground-controlled commercial aircraft that was destined for a crash landing and fuel additive testing was carried out in the 1980s. While the fiery impact images and videos have been publicised over the years that followed, the control systems allowing for numerous remote take-offs and landings were taken for granted by the wider public.

The commencement of the Gulf War - Operation Desert Storm - in 1990 saw the US military using remotely operated aircraft carrying video cameras and sensing equipment. Videos taken by the drones showed the opposition forces retreating or regrouping, and the public started to appreciate the uses of an unarmed drone. After the conflict ceased, armed drones were introduced into operational military conflicts and publicised like never before, for example, the post-9/11 operations in Afghanistan and the subsequent Iraq war.

Both Airbus and Boeing as major aircraft manufactures made an astonishing admission in 2003 to the worlds' media in very broad general details. Both manufacturers stated that they were close to perfecting the technologies required to remotely land a civilian passenger aircraft that was experiencing a 'hijack' type event. Boeing submitted a patent and received confirmation of this intellectual property in 2006. However, this new concept did little to avert the various acts of pilot homicide that occurred in the years that followed. The technologies to automatically land an aircraft have been fitted to all large commercial aircraft and are based on the airports Instrument Landing System. Such systems are so advanced that the aircraft can land automatically in zero visibility on a runway and stop – while the pilots observe the computers and autopilots managing this task. Thus, the technologies to remotely command an aircraft to land within an 'uninterrupted landing system' exist, with minimal additional equipment needed to achieve this.

Other considerations are explored with respect to commercial aircraft and their current weaknesses. These include the lack of mechanical keys/codes required to start a $300 million USD aircraft, the ongoing risks posed by in-flight fire, or access to the electrical and avionic bays. Likewise, consideration is given to aircraft that are commanded to land by remote control from the manufacturer, and the question is posed as to whether the flight crew should be passive at this time. Also discussed is how further acts of sabotage can be prevented, allowing for a safe landing and positive outcome. In addition, modifications to the critical life support systems, electrical generation systems and electronic communication equipment (including the transponder) are proposed, as it remains questionable that commercial pilots are able to disable critical systems from the protection of their locked flight decks.

Lastly, the justification and drive for modifications to commercial passenger aircraft (that are foreseen) will be attributed to financial savings that can be made from a fully data streaming aircraft. Reductions in fuel and maintenance costs will be directly attributed to a satellite streamed data service that monitors all of the aircraft's systems. Financial savings by the operators (the airlines) would drive this process, and while the levels of security would improve, the reductions in both cost and environmental emissions would be the predominant factor in making these changes.

9.2 History of Remotely Controlled Aircraft

As with any large, protracted modern military conflict, aviation technology developed and new applications for the devices were found. During the 1914–18 First World War (WWI), pilotless aerial vehicles were developed, and novel applications were conceived. In 1918, the US military considered methods of using new aircraft technologies with explosive ordinance, resulting in the `Kettering Aerial Torpedo'. The project was the combination of Charles Kettering, an electrical engineer; Orville Wright as a consultant from

FIGURE 9.1
US military's secret Kettering Bug drone. (USAF.)

the Dayton-Wright Airplane company; and Elmer Ambrose Sperry, who designed the control and the guidance systems. The aircraft had a maximum range of 75 miles, and the device could deliver around 180 pounds of high explosive to the target. A photograph of the drone is shown in Figure 9.1 - note the device launched from a dolly and track system. Due to the end of the war, the device did not see active service, yet its development and use remained a closely guarded secret for many years to come.

In 1935, the British military developed a radio-controlled drone to be used for target training, called the De Havilland DH.82B Queen Bee. This aircraft was a radio-controlled variant of the De Havilland Tiger Moth biplane. Different variants of the Queen Bee were derived, including a seaplane, model L. A photograph of the L5984 (Figure 9.2), taken by the British War office shows the aircraft on the launch ramp, with Prime Minister Winston Churchill in the foreground.

In 1946, the American military modified a B-17 Flying Fortress aircraft (see Figure 9.3). The remote-controlled drone took-off from Hilo Naval Station in Hawaii (August 1946), flying 2,600 miles to Muroc Army Airfield in California. The aircraft was controlled by Army Airforce personnel from the Muroc Army Airfield in the USA, who remotely flew the aircraft for the 15-hour flight.

FIGURE 9.2
Queen Bee seaplane with Prime Minister Winston Churchill. (British War Office.)

FIGURE 9.3
B-17 remote-controlled aircraft flying from Hawaii to California on a 15-hour endurance flight,
August 1941 (Imperial War Museum). (USAF.)

The use of radio technology for more complex aircraft that were to be intended as target drones was expanded in the years that followed. The UK's English Electric Canberra was the first medium-range jet turbine-powered high-altitude nuclear bomber. In 1957, the reconnaissance variant flew to an altitude of 70,310 ft, breaking flight records. The versatile application of radio control technology allowed for the production of the Canberra drone series (including for the USAF as the Martin B57A), where the military would use new high altitude guided missiles to shoot down this high altitude performance drone (Figure 9.4).

The use of unmanned drones for target practice had become a common use of this electronic technology. A new application on the battlefield was the use of a ground-controlled drone that could carry reconnaissance equipment to transmit back the location of battlefield targets while flying over hostile territory.

The USAF extended this technology to launch test drones from C130 aircraft flying at altitude (illustrated in Figure 9.5).

FIGURE 9.4
Martin B57A/Canberra target drone. (USAF.)

FIGURE 9.5
C130 aircraft carrying various drones in-flight, prior to their launch. (USAF.)

9.3 Federal Aviation Administrations' Full-Scale Controlled Impact Demonstration

There have been numerous reconnaissance drones designed and produced for military applications in the subsequent years, with increasing levels of technology and autonomy. A civil application for remotely flown large commercial aircraft was instigated by the Federal Aviation Administration, USA, in July 1980. It is known that when a large passenger aircraft crashes into the ground during a forced landing, the fuel tanks that are contained within the wings tend to fail, spilling the kerosene fuel onto the ground. The objective of the full-scale crash test was to fly and deliberately crash on landing an old commercial aircraft. The overall aim of the full-scale crash was to evaluate a new fuel additive, with the hope that the post-crash fire could be mitigated by using additives. The aircraft selected for the final crash test was a Boeing

FIGURE 9.6
Converted Boeing 720 passenger aircraft – practicing a climb away during the final approach in RPV conditions. (FAA.)

720, with the electronics and guidance control systems being provided by the National Aeronautics and Space Administration's Ames Research Center, USA. NASA was seen as the experienced partner with considerable expertise in Remotely Piloted Vehicles, (RPV) with Flight Control Systems. Furthermore, NASA had proven the reliability of this technology *'in the early 1970s as a means of flight testing experimental aircraft and advanced technologies in a far less hazardous manner'* (FAA Full-scale Transportation Controlled Impact Demonstration Program, September 1987). The aircraft was modified to be able to be operated fully remotely, and during the project this B720 aircraft conducted 69 separate flights; 9 fully Remotely Piloted Vehicle (RPV) take-offs; and 13 RVP landings (see Figure 9.6).

The controlled impact into the runway, as per the experimental specification, did not fully go to plan, but the crash test (see Figure 9.7) provided valuable safety data and information to the industry. Figure 9.7 shows the controlled impact with the ground, and the unexpected yawing that followed as the aircraft slid across the surface.

This project highlighted the capabilities of remotely piloted flights (controlled from the ground) of this large, complex multi-engine turbine aircraft, using the state of the art technologies in the early 1980s.

FIGURE 9.7
Converted Boeing 720 passenger aircraft – practicing a climb away during the final approach in RPV conditions. (FAA.)

9.4 Remote-controlled Aircraft (Drones) During and After the Gulf War – Operation Desert Storm

The summer of 1990 was a pivotal moment for the Western military powers. In August, Iraq made a surprise invasion of the small state of Kuwait, taking the entire neighbouring territory with ease. The western governments formed an alliance known as Desert Shield from August 1990 to the end of February 1991. In those months, significant preparations were made to retake the occupied Kuwaiti territories, but the gathering of live aerial images and data was deemed to be high risk. The US military deployed numerous unmanned drones, that were not armed, to survey the area. They fed back tactical data for Operation Desert Storm's active military phase. Images from

FIGURE 9.8
Pioneer RQ 2B unmanned aerial vehicle. (Smithsonian Museum.)

the drones were released to the world's media during Desert Storm, illustrating the superior use of the '*new*' drone technology. This was the first time that a military had extensively released images from drones that were telling the story of the progress that was ensuing. One drone publicised was the Pioneer RQ-2A unmanned aerial vehicle (Figure 9.8), which could be flown from a remote land station using modern communication means. The video clips shown at the time in these war briefings included the extensive use of the long-range video optics, filming the hostile forces going about their duties before and during the conflict. During the war, several military pilots were shot down over hostile territories, and the Iraqi government paraded these prisoners of war on their state TV. What became apparent to the media from these images were the capabilities of a remotely flown aerial vehicle, which carried no persons and posed little risk of the loss of life should they be shot down and lost. The concept of expanding the capability of drones, to carry weapons now became publicly acceptable, to which many militaries and manufacturers recognised.

In the years that followed, new armed variants of unmanned aerial vehicles were developed, and perhaps the most iconic of these drones was the General Atomics Predator (Figure 9.9), which was initially based on an unarmed variant, but was further developed to be armed with Hellfire

FIGURE 9.9
General Atomics Predator (Grey Eagle) unmanned aerial vehicle.

missiles. The regular deployment and use of these armed drones became frequent, after the 9/11 attacks and subsequent missions in the Afghanistan region. These drones have an automated take-off and landing capability, in addition to modern guidance avionics, that has similar equipment capabilities to modern aircraft, thus allowing their ground crews to manage the workload effectively.

A much larger drone that was developed around the same time was the Northrop Grumman RG-4 Global Hawk surveillance drone (Figure 9.10). This UAV is designed to carry numerous sensors including radar surveillance systems based on U2 platforms, but the sheer size of the aircraft makes it noteworthy because the Global Hawk has a similar-sized wingspan to a B737 aircraft. In 2001, the Global Hawk UAV became the first aircraft to fly non-stop, making a fully autonomous flight from the USA to Australia, some 23 hours in duration.

While these long-distance flights of unmanned vehicles do not appear to be particularly exciting or challenging, their routine nature has allowed the technologies for flight control, guidance and semi-automated operation to become a 'mature' discipline, with vehicles now given permission to fly in the national airspace, i.e. not being excluded from commercial traffic.

FIGURE 9.10
Northrop Grumman RG-4 Global Hawk unmanned aerial vehicle.

9.5 Boeing Patent on Remote Control Takeover of Aircraft

The Boeing Company submitted a patent (US714971B2) to the US patent office in February 2003, with the grant of the patent in November 2006. The application that was submitted by Boeing details that the countermeasures for commercial aircraft flight deck security (e.g. hijack) are not perfect, and although armed Sky Marshalls may be present onboard passenger aircraft, there could be times when the presence of a reinforced flight deck door would be counterproductive to the safety and security of the aircraft and passengers. The patent description details the need for an uninterruptable (control) landing system that can be activated from the ground or in the air, to be able to take over the control of the aircraft to land at a designated airfield, taking over from the occupants of the aircraft. The system foreseen is said to be *uninterruptable*, meaning the Automatic Control System 'control box' that interfaces with all the aircraft systems would need to have its own power source and supply, and that once it has been activated, the aircraft must land. Upon landing, the system would require deactivation by approved ground service engineers, thus taking off again after a 'forced landing' would not be possible without the manufacturer's agreement. Boeing cites that this technology concept could be applied not just to aircraft undergoing a hijack scenario, but also to boats, ships, trains, buses, etc.

The uninterrupted navigation systems are believed to use additional guidance chip-based technologies that were previously developed for cruise missiles. The new navigation systems are believed to use processors with built-in gyro accelerometers; thus, the inclusion of three of these processor chips will give the same navigational capability as a ring laser gyro, which forms the current commercial aircraft Instrument Navigation System. In addition to receiving satellite data, the aircraft uses GPS data from the enhanced Ground Proximity Warning Systems (GPWS). Thus, the aircraft `calculates' its geographical location (and speed, altitude, acceleration in all axis, etc.) from the space-based GPS systems, receiving a minimum of four standardised time signals from geostationary navigation satellites. Meanwhile, the enhanced GPWS system alerts pilots to the proximity of terrain, based on the combination of the latitude/longitude and ground elevation database that is stored within the unit.

Another important automated system that modern aircraft have relied on over the years is the Autoland system. These Autoland technologies allow aircraft to approach an airport, descend and finally land. The autonomous landing capability is an old radio technology that is very mature, with the first fully automated landing system being trialled in 1964, UK. The destination airfield transmits two distinct radio signals (a localiser and a glide slope radio signal) that the approaching aircraft receives, indicating whether the approach trajectory is *correct*, *too high* or *too low* concerning the 3-degree approach glide slope, and *left* or *right* of the centre line of the runway. Current commercial aircraft are fitted with these Instrument Landing System (ILS) capabilities and combine this mature technology with the Radio Altimeter (RA) on the aircraft. The RA accurately measures the altitude of the aircraft using radio waves up to 2,500 ft from the ground: it is highly effective and fitted to every large commercial aircraft. The most current level of approved Autoland system is an ILS Category III C system. The ILS Cat III C allows the aircraft to approach an airport in *zero visibility* weather using the autopilot. Furthermore the autopilot, when the system is active can fly the aircraft using the ILS data to provide navigation corrections to the flight controls all the way down to the runway. This controlled, full-precision landing will also deploy the spoilers on touchdown and apply the brakes to bring the aircraft to a full stop. This means in a CAT III C landing, the pilots might not have even seen the ground prior to landing, because the need for a Decision Height is removed from the approach requirements.

The combination of satellite positioning, the additional gyro navigation system and fully matured Autoland system is believed to be the backbone of the navigation system for this uninterruptable landing system.

The Boeing Aircraft Company made a number of public announcements and press releases to the worlds' media in 2006 regarding the principles of this patent and claims of the technology, yet there is almost nothing published in the open literature or Boeing technical publications (i.e. Aircraft Maintenance Manuals (AMM), Flight Crew Operation Manuals (FCOM), etc.). Furthermore, since the application and grant of the patent, there have still been a number of recorded homicide (or potential homicide) events, and if an aircraft lands at a designated airfield due to a security event, these are widely reported. For example, on 6 February 2000, a Boeing B727 belonging to Ariana Airlines, Afghanistan, was hijacked on a domestic flight planned flight from Kabul, Afghanistan. The aircraft made two fuel stops in Asia, a fuel stop in Moscow Russia, before attempting to fly to London Heathrow, UK. The UK's designated airfield for security is Stansted Airport, some 35 miles northeast of London's city centre.

Note: it is important for aircraft undergoing security events not to overfly densely populated metropolitan areas, and Stansted Airport meets that requirement. On approach to the UK, the hijacked aircraft was met by the Royal Air Forces' Quick Reaction Alert interceptor fighter planes who accompanied the aircraft, as per usual. Upon landing, the aircraft was parked at the 'designated area' for such events, away from the commercial passenger buildings, to the west of the active runway and the terminals (adjacent to the FLS maintenance hangers and Inflite VIP reception) and a tense stand-off commenced. As news of the hijack became known to the world's media, the various outlets dispatched reporters and photographers to follow and record the unfolding events (map shown as Figure 9.11), with 180 passengers and seven crew being held by the perpetrators. After six

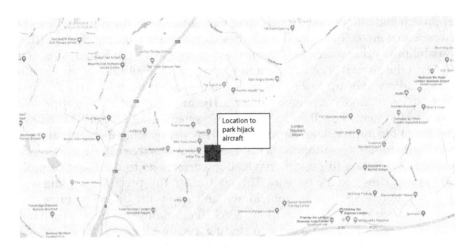

FIGURE 9.11
Map of Stansted Airport and the illustrated location of hijacked B727 – parked, relative to the terminal building. (Google Maps.)

days of captivity, the UK's security services negotiated the eventual release of all passengers and crews. More importantly, this was done without the need to deploy the Special Air Services (SAS) from the British Military. The use of the SAS to potentially storm the aircraft was a real consideration, as the military had flown in a near-identical B727 on the morning of the 6 February to a Stansted airport hangar, to continuously rehearse counter-terrorism activities using live ammunition. Six days of live shooting at this identical B727 aircraft resulted in six weeks subsequent maintenance to repair all the bullet holes.

Boeing does not have the exclusivity on the principle of remotely pilot-ing a commercial aircraft to provide an uninterruptable landing system. In December 2003, Airbus (EADS) made a press release to the media, together with Honeywell electronics and BAe Systems, stating that they were '*close to perfecting technologies that takes control of airplanes to prevent them crashing into obstacles.*' The press release suggested that trials on smaller aircraft had been conducted (successfully) but admitted that *the idea of completely turning an airplane's controls over to a computer could make people nervous.* Furthermore, in June 2020, Airbus announced that their two-year Autonomous Taxi, Take-off and Landing (ATTOL) project, with automated flights in their A350 air-craft, had been completed successfully. An A350 test aircraft is shown in Figure 9.12. This ATTOL project included more than 500 flight tests, includ-ing six final flight tests and five take-offs and landings, which have been used to prove the system operations and machine learning of the control algorithms. The Airbus ATTOL project is based on the previous findings from *Project Wayfinder*, which is also known as *Acubed*. *Project Wayfinder* used

FIGURE 9.12
Airbus modern A350 on test flight. (Antonio Velasco Cruz.)

high-quality optical images on small, electric vertical take-off or landing demonstrator aircraft, 'to see' and interpret the layout of the runway/taxiways, etc., in much the same way a Tesla electric car does on land using the autopilot function.

9.6 Current Capabilities and Their Limitations

The advancement of unmanned aerial vehicle technologies and operations that are presented thus far clearly demonstrate that the remote control of a commercial aircraft from the ground is achievable with today's technologies. Likewise, the autonomous navigation of an aircraft experiencing an unforeseen security event (e.g. hijack) is based on proven military systems from cruise missiles. Lastly, the automatic landing systems using the ILS airfield broadcasts, which are integrated into the autopilot system *are the final piece in the puzzle* for a ground-based uninterruptable landing system. All the necessary technological elements for autonomous flight have been proven over the years, and the admission by both manufacturers that they have been working on such take over systems is interesting.

However, in-flight aviation security events that are met by interceptor military aircraft are reasonably frequent events. While they do not appear to take place every month, their frequency is recorded because when they do take place, the world's media reports all the occurrences in detail. The widespread deployment of a working 'uninterruptable landing system' on all commercial aircraft is highly questionable because so many high-level occurrences have taken place (involving or believed to involve pilot homicide/hijack) where the final outcome was not 'positive'.

For example, the Airbus A320 on 24 March 2015 Germanwings Flight 9525:, when F/O Lubitz locked the Captain out of the flight deck using the cockpit door toggle switch (Figure 8.22), this should have transmitted a secure data message to Airbus Toulouse that an unusual event was taking place (in real time). Furthermore, the autopilot selection to descend the aircraft to +100 ft, while flying at high altitude over mountainous terrain, should have been sufficient to activate such a ground-controlled uninterruptable landing system. This is assuming that a covert system by Airbus, BAe Systems and Honeywell was installed on this A320 aircraft that was manufactured in 1990. Clearly, such a system was not installed, operational or effective, because the annihilation of the passengers, crew and aircraft in the ground impact enroute indicates that the F/O had full control of this aircraft until the bitter end.

Likewise, the final Malaysian accident report for the MH370 flight acknowledges the Boeing uninterruptable landing system that could be fitted to aircraft, but implies that as the original manufacture date of 29 May 2002 was before the grant of Boeing's patent in 2006, such a system was not present or installed.

Another explanation that might explain why an uninterruptable landing system was not activated for the MH370 or Germanwings 9525 could be the system is a subscription-based product. The MH370 event highlighted that the operator did not use the satellite communication subscription for the maintenance applications. If Malaysian Airways were not willing to subscribe to an engineering system that would reduce their overall costs associated with operating an aircraft under normal conditions, they would not subscribe to a covert landing system. The plausibility that Boeing would offer the uninterruptable landing system as an option is based on the Boeing corporate strategy further to the Boeing 737-Maxx crash events and fleet groundings. The B737-Maxx series requires an automated nose pitch system to be fitted, due to the change in the engines and the nose-up pitching in high power settings. Unfortunately for Boeing, the safety system fitted by Boeing relied on a single Angle Of Attack (AOA) sensor, which could be subjected to damage and would activate a pitch down attitude. Boeing offered the airlines an upgrade modification from one sensor to two sensors, and also an indicator inside the flight deck to indicate disparity between the two AOA sensors. Why Boeing considered the installation of the stall recovery system's dual AOA sensors as optional has yet to be explained: T-tail aircraft use an almost identical AOA sensing system, and they have been fitted with two AOA sensors for the last 50 years. It would appear that Boeing, as a major manufacturer, considered some systems to be revenue stream, and not part of the core product.

9.7 Changes and Technologies Required for a Safe Autonomous System

Reviewing the published pilot homicide events, the underlying findings of the subsequent investigations indicate that the commercial flight crews have a full understanding of their respective aircraft, including the vulnerabilities and limitations of all the technical systems. All the life support system controls are located inside the reinforced flight deck. While the manufacturers may have intended pilots to have the final decision as to disabling such systems, historical events indicate that some systems should not be switched-off by a flight deck occupant without first declaring an emergency, as the continued support of human life aboard the aircraft is dependent on the Environmental Control System functioning correctly. Likewise, the full electrical systems (including the communication systems and transponder) are all controlled and isolated from within the flight deck.

9.7.1 ECS Life Support

For such critical systems that affect life support and the normal powered operation of the aircraft, the justification for switching-off such systems in-flight is highly questionable. In extreme cases, such as a double thermal event on the aircraft air-conditioning packs (forming part of the ECS), in addition to making an immediate unscheduled landing, it would be prudent for the manufacturers to change the procedure for disabling such systems in-flight. The current protocol is for the manufacturer to cover the switch with a spring-loaded wire cage – but that can be operated by a single person inside the flight deck. If, however, two separate and unique alpha-numeric codes were required to be entered by both pilots into the flight computer in order to isolate these systems (when the aircraft was weight off wheels in a flight condition), this would mitigate the risk of deliberate subversion. Ironically, as the ECS is fully computer-controlled, another viable solution could be modification of the ECS computer software to fully remove the pilots from the in-flight 'isolation' controls, and only allow the manufacturers via the satellite uplink to securely make any necessary system isolations. The same protocol should be applied to the electrical systems – requiring both pilots to input a unique personal code into the aircraft, or having the manufacturer take control of the systems via a secure satellite uplink.

9.7.2 Circuit Breakers

Currently, the circuit breakers (C/Bs) for all the aircraft's electrical power connections are located either on the overhead panel (above the centre pedestal), the general flight deck compartment side panels or inside the electronic and electrical bay (usually underneath the flight deck). All these circuit breakers can be isolated by pilots in-flight, but the location being within easy reach is based on the pilots' historic troubleshooting abilities to work around problems. While this 'can do' strategy might have been highly advantageous in the post-Second World War (WWII) years, the current complex aircraft often require advanced computing to identify and rectify technical problems. Many of these technical diagnoses and rectification tasks are performed by highly trained qualified avionic licensed engineers. The justification to have all the C/Bs within easy reach of pilots, giving the flight deck occupants the ability to switch off all electrical power in-flight based on the pilot's wishes, cannot continue to be justified in light of various homicide events. One possible solution would be to relocate the current C/Bs to an unpressurised part of the aircraft, removing the possibility of in-flight resetting or opening. If a C/B does open due to operational difficulties, then a decision should be made by the airline's maintenance team (with input from the pilots) whether to land at a nearby airport to perform an unscheduled maintenance task, or to press on to the final destination.

9.7.3 Transponders

The transponder operation likewise requires some modification. The current process is to allow flight crews to decide (selecting the operation using a rotary switch) when to switch the system *on* and *off*. The events surrounding MH370 have indicated that having such a capability is a significant security weakness for the industry. A simple modification would be to change the operation logic in the computer-based controls. Once the transponder is set with the four-digit code (on the ground), the aircraft takes off, and it should not be possible what-so-ever to switch off the system. Deactivation of the transponder should be automatically performed when the aircraft has landed (weight on wheels), the ground speed (from GPS data) is shown to be zero and the park brake is set to *on*. Only then should the transponder Mode S cease transmissions.

9.7.4 One Time Use Codes

Aviation and aircraft security are complacent as an industry. Small general aviation aircraft require the occupant to use a metal key to be able to start the aircraft's piston engine, as the aircraft has a new retail value up to $750,000 USD – depending on the size, avionics, etc. Perversely, a new B787 is around $250,000,000 USD and a B777 320,000,000 USD, yet large heavy commercial aircraft have no keys, no security codes to start the engines. The theft of Horizon Air's aircraft by a ground staff member demonstrates this fact very aptly. Airports rely on deterrents in the form of fences, razor wire, cameras and patrols. Yet, these defences are regularly tested and breached by stowaways, members of the public who have no hostile intent, etc. Ironically, the security levels applied to the travelling public are excellent, i.e. the security screening inside the passenger terminal. A computing-based solution is necessary, to provide a one-time code that contains 'all the information necessary' for the aircraft to be operated correctly. For example, if an aircraft undergoing maintenance requires a ground run, the maintenance base should be able to issue a code to an engineer that would allow for the engine to be switched on and operated. If the system detects that the ground speed of the aircraft (from GPS data) is greater than the maximum taxi speed, the aircraft computer should automatically cut the fuel to the engines. For commercial flights, the expected route including waypoints, intended airways, etc., should be inputted into the flight planning software as normal. The one-time use code is then generated by the flight planning software in conjunction with the OEM server and ATC flight planning submission, issued to the pilots and used to start the engines after pushback from the terminal. If during the flight the pilots significantly deviate from the filed flight plan, this is automatically brought to the attention of both ATC and the airline. One-time use codes such as this are standard use by freight forwarding companies, allowing customers to track their shipments. Assuming the aircraft's Flight Management Computer (FMC) has Wi-Fi access or Satcom data when parked at the gate, the single-use code could also be used to download the flight's full route into the FMC. This integrated approach would prevent pilots from flying the incorrect sector which has happened in recent years on more than one occurrence.

9.7.5 Satellite Communication Uplinked Continuous Data

Many commercial aircraft are fitted with satellite communication technologies. Some airlines use the data communication capability to communicate between the aircraft and the airline (via the OEM's data server), such as the Airbus Skywise service. Newer aircraft also have the possibility to use the Satellite data transmission for up-linking the ADS-B aircraft performance data (including the transponder information) to ATC, etc. However, the use of such services has financial implications associated with it the subscription and transmission/receipt of data from/to the aircraft, and the subscription to the OEM's maintenance software. It is feasible for the operational data of the aircraft to be compressed into very small packets (of data) that can be uplinked via the satellite communications, so the information is more compact and efficient. If all the data that is stored in the DFDRs and DCVRs was compressed and continuously uploaded via satellites to a cloud-based service, in time the regulators might allow for the final removal of 'black boxes', because the final moments of the aircraft will be available.

With mature micro-sized charge-coupled devices (CCD) forming optical capture devices (e.g. mobile phone front and rear cameras), it is now possible to embed numerous streamed CCD images to a cloud service. The flight deck and entrance from the cabin to the flight deck would be the prime locations, in addition to the external positioning that many aircraft already have (e.g. B777-300 with nose gear camera, wingtip views from top of vertical stabiliser). The inclusion of the external coverage of the aircraft would give a safer ground manoeuvring capability, reducing the possibility of unexpected aircraft/equipment/vehicles striking the aircraft. Likewise, for in-flight operations, the opportunity to view a turbine engine unexpectedly discharging fluids from one of the drain masts, or a partially extended main landing gear leg would allow for more precise performance-based decisions to be made. Such CCD devices are so small, inexpensive and light, the justification not to fit them on the grounds of saving weight or cost is not a valid argument.

9.8 The Justification and Driver to Introduce Ground Monitored Technologies

The world's aviation businesses are based upon the carriers (i.e. the airlines) being able to transport revenue, passengers and freight, with the principle objective for the airline business to be profitable. In the purest and most simplistic terms, if an airline fails to make sufficient profits, then bankruptcy is guaranteed. From a passenger's perspective, they identify the route they wish to fly from, the destination, the time/date of proposed travel and the class of travel (economy to first class). Choices for travel are based on the total travel time, the route and most importantly, the total price of the flight ticket(s). Most travellers are sensitive to price in the selection of the flight. In Europe and North America, all the airlines are very carefully regulated and monitored by the national aviation authorities: minimum maintenance standards are maintained, and the overall level of safety is incredibly high in terms of extremely low accident/incident rate. Statistics indicate that the passengers travelling from home to the airport are significantly more likely to be involved in an accident in the surface transportation phases versus the flying phase. The travelling public are savvy to the aviation accident rate, and their expectation is that the flight they will travel on will be routine and free from problems or risk, all underpinned by the extremely low fatal accident rate. This expectation is based on past personal experience.

9.8.1 Financial Drivers

Chapter 6 discussed the human factors and safety management system implications. The improvements in the late 1990s cited were based on the principles that not making errors would lead to more reliable aircraft; consequently a 3% financial gain can be simply made in the Engineering annual budget. A similar approach to financial savings must be used to underpin any future aircraft modifications and changes to the procedures. The travelling public will not be willing to pay an additional anti-terrorism or anti-pilot homicide premium on each and every flight, because the frequency of such events is so low. That said, the security risk cannot be discarded, as repeated historical events have proven, as the most intent individuals will plan and carry out their dastardly acts, destroying aircraft and killing innocent passengers and crews.

The key to any potential improvements that have been suggested thus far in this chapter must be based upon the ultra-long-term financial savings that airlines seek, and manufacturers claim. All modifications will need to be driven by the Airframes' Original Equipment Manufacturer, e.g. Airbus or Boeing. Changing the operation of the aircraft to limit what can be done in-flight by flight deck occupants is a task that can only be achieved by the airframe OEMs. Chapter 5 presented the background and state of the art technology for maintenance management. While some data are transmitted live, much of the aircraft's performance data are downloaded on a weekly basis, with each download providing around 1 TB per sector. If all such data were to be uploaded via satellites, the quality of the aircraft operation and maintenance predictions would improve. While the carriers will be concerned that the transmission of aircraft performance data is expensive, the irony is that major social media platforms are currently working on a low earth orbit provision to allow users to upload their activities/photos/videos using an ultra-low-cost satellite service. This is because much of the world's social media target demographic audience does not have an effective 3G/4G/5G mobile telephone provision, or a cable-supported Wi-Fi access. The development of this type of low-cost social media satellite data service would be a disruptive product, and the traditional satellite data providers would need to adjust their data pricing accordingly to compete with the new emerging technology. These low-cost satellite communication services have been in development for a significant number of years, with the large corporations behind these future services being Facebook, SpaceX and OneWeb.

9.8.2 Live Streamed Data Reducing the Fuel Burn

The potential gains from a full operational live data stream, with the introduction of the single-use code per sector (to start the engines/operate each sector) would also bring a second measurable gain: namely that of routing and efficiency. Currently, while routing is planned within the Flight Operations of an airline, the active flight crews can request ATC to vector to different bearings/altitudes as they see fit. An active system would allow the ground staff to communicate better with flight crews when 'more favourable' flight profiles are considered. ATC would benefit from such a system. This is because ATC's electronic planning infrastructure would be able to much more accurately predict when the airspace would be occupied by the respective aircraft, so the most optimum applications can be applied. For example, if saving time was the principle factor, aircraft on approach could be given a higher approach speed before commencing the 'long final approach'. Alternatively, a more environmental approach would be possible with Continual Descent Approaches (CDAs), allowing the engines to use significantly less fuel from the top of the descent, converting the potential energy gained from altitude into the kinetic energy of the airspeed in this steady descent profile. Furthermore, the streamed live performance data would quantify the actual fuel burn in real time, and as fuel consumption has financial implications, this would be an immediate gain that the airlines would like to monitor more carefully. With reduction in fuel consumption, there comes a significant emission reduction capability, because the output of each engine can be recorded, optimised and compared. The data would contain the values of the reduced fuel flows, the reduced lubrication oil consumption, and both of these performance metrics can then be employed to calculate the reduction in the CO_2 and changes in other gaseous emissions (e.g. NO_x, NO, water vapour, etc.). These accurate, logged savings in environmental gains (and reduced fuel burns) could be used to reduce the costs of the EU's Emissions Trading Systems (EU ETS), a blanket charging system on all large commercial aircraft entering and operating within the European airspace. The data would support an immediate EU ETS cost saving that is imposed on the carrier.

9.8.3 Live Streamed Data Reducing Deviation from Flight Plans and Further Reducing Fuel Burn

The ability to monitor the flight crew's adherence to the approved flight plan in a live-streamed data context would also highlight any changes in the human performance (e.g. latencies). Fatigue in flight crews has always been an underlying problem that airlines and regulators have attempted to monitor and improve. Unless the observer is within the flight deck, the use of pilot-targeted questionnaires using Likert scales to estimate levels of fatigue and lethargy is not accurate, and lacks numerical quantification. The live data streamed information would allow much more sensitive observations to be made and recorded, perhaps with comparisons to an individual's previous flight performance data sets, or with the performance mean of the fleet based crews. The fewer deviations made by the pilots flying the plan, and the increase in the monitored flight characteristics would potentially reduce the fuel burn and improve the environmental aspects.

9.8.4 Data Security and Encryption for Uninterrupted Landing Systems

If an uninterrupted landing system were to be fitted to a live commercial aircraft, the principle objective would be for such a system to be sufficiently secure from interference from the non-authorised sources. Data encryption is a highly mathematical technique that allows for packets of data to be transmitted and received, preventing the information contained within the transmission to be decoded or changed. Hackers, malicious programmers and rogue states would have an interest in taking command of a distant commercial aircraft, and potentially causing a homicide act by deliberately crashing the aircraft into the ground/other aircraft/building/sea. Therefore, any system fitted would require the airframe OEM to have operational military-grade encryption, such as Advanced Encryption Standards (AES) using a 256 bit-key or higher with an End 2 End Encryption (E2EE). Military communications require the highest levels of security. For example, a government with a nuclear deterrent wishes to allow its leaders to control the nuclear devices, thus assuring total security. The political leader requires the ability to transmit coded signals and data transmissions to submarines, etc., commanding a nuclear attack if needed, as detailed by the Ministry of Defence/Defence Nuclear Organisation, UK. If such communication levels were compromised, such is the risk, it would be possible to deploy and detonate a nuclear device without even needing to be physically present with such a nuclear weapon. Military encryption is always evolving, including the application of quantum physics, where pairs of electrons can be used to code and transmit data, ensuring that data cannot be intercepted or deciphered without the intended recipients being made aware of the security breach.

The Boeing Corporation, Airbus Group, BAe Systems, etc., are all providers of the most secure data systems to the militaries in the USA and Europe. As with any such system that has been designed, endorsed and observed by any of these major airframe manufacturers, the expectation is that the highest level of AES would be employed, and not the lowest cost option that has recently come to light with one manufacturer's safety systems. Any uninterrupted landing system would need to update its capabilities remotely, in a similar manner to the way personal computing operating systems communicate with their designers to perform regular security patch updates.

The Flight Management Computer and Satellite Data Unit would be the ideal avionic computing systems to host the additional circuit boards within these Line Replaceable Units (LRUs). The inclusion of a bespoke printed circuit board in both LRUs would give the necessary performance requirements, allowing for fully-encrypted data to be received from the OEM, and aircraft performance to be transmitted securely back to the OEM servers.

9.8.5 Uninterrupted Landings and Risks Posed from the Aircraft's Occupants

If a commercial aircraft were to require the deployment of the uninterrupted landing system, one significant risk that is posed is the persons that remain inside the flight deck and the passenger cabin. It can be assumed that hostile persons who have taken command of a commercial aircraft, which subsequently has an uninterrupted landing system activated, are not going to remain passive observers when the aircraft descends and lands under the control of the manufacturer. Chapter 7 highlighted the dangers presented by fire in a confined space, such as a pressurised aircraft. The flawed concept of flight crews switching-off the aircraft's ventilation system in the vain hope of extinguishing a fire is a historical decision that has proven to kill the passenger. These occupants would suffer the immediate effects of smoke inhalation because the gases are highly toxic, and the reduced ventilation flow has little or no effect on the seat of the fire. Worse still, if the aircraft is flying at altitude (about 20,000ft), the loss of pressure encourages the onset of hypoxia in the aircraft' occupants. A fire that becomes established with a high heat flux, with the exposed surfaces being heated by radiant sources to temperatures exceeding 500°C (932°F) will cause a flashover event, where the pyrolosis of materials in the zone will produce a flammable gas that will spontaneously combust. If the occupants within the aircraft recognise that the aircraft is no longer under their direct control, and they were to start a fire, it is plausible that before the aircraft can land the fire's flashover will cause the aircraft to crash catastrophically – something which an uninterrupted landing system has no direct control over. Fire events are not a new problem for airlines, and the UK's AAIB final report (1988) for the British Airtours engine fire at Manchester airport in 1985 directly addressed the problems pertaining to fire in an aircraft that leads to mass fatalities. The report findings specifically recommend that the airlines provide smoke hoods for all of the passengers, and the manufacturers modify the aircraft to include a water spray system (drawn from the potable water supply) as the technology existed in 1988, and would form a lifesaving 'twin strategy'.

Another potential source of aircraft in-flight sabotage during the descent phase would be posed if an individual gained access into the electrical and avionic bay. As these bays are located within the pressurised areas of the airframe, an occupant can easily access this sensitive area and interfere with all of the equipment. A solution to this significant problem would be to retrofit

FIGURE 9.13
Airbus A380 photo with Electrical and Avionic bay proximity to the forward cargo bay door-rastered boxes.

the floor trapdoor from the flight deck to the bay area with the same protective measures used on the flight deck door. Multiple electronic locks would be necessary; the structure around the trap door would be increased; and all the floor panels from the passenger cabin forward, including the flight deck would need to have ballistic Kevlar materials (included in the composite structure) to prevent ballistic penetration or attack. The panels will also require fastening securely from below the cabin structure, to prevent a determined passenger/pilot, etc., from lifting up the floor panels in the cabin to gain access to the hold and the sensitive avionic bay. This is because the electrical and avionic bay protrudes aft of the flight deck, behind door 1. The checkerboard rastered boxes (Figure 9.13) on the nose of the aircraft, represents the electrical and avionic bays. It is possible to open the forward cargo bay door (from the outside), and inside the forward cargo bay (on the right or forward bulkhead) is a small door. This small internal door (which can be opened from either side) leads into the electrical and avionic bay. Once inside the electronic and avionic bay, an individual can climb up a small ladder, open the floors' trapdoor and climb up to enter the inside the secure flight deck (behind the observers' seat). A simple internet-based search will identify videos of pilots demonstrating the navigation of this route, from the flight deck down to the cargo bay, returning to the flight deck.

Likewise, airlines will need to give careful thought to aircraft fitted with underfloor rest areas for crews, or underfloor toilet facilities. Both of these features would give a determined cabin occupant access to the forward cargo compartment and latterly the avionics bays and the 'secure' flight deck. Modifications would require the additional use of further Kevlar in composite wall panels, etc.

Lastly, any uninterrupted landing systems fitted and monitored by the manufacturer must not be considered by the OEM to be a subscription-based service. The OEM should consider the costing of any potential system as a positive marketing for brand, to make their airframe the safest and most efficient in its class. The airline's potential cost savings will drive the sales and leasing of the aircraft that are fitted with full monitoring and communication devices.

9.9 Conclusions

The history of remotely controlled aircraft has shown that the concept is almost as old as the first powered aircraft. The milestones over the years, including post-WWII are considerable, but most of the developments took place under the cover of secrecy and military superiority. Occasional civil aviation testing has emerged, such as the Federal Aviation Administrations' full-scale controlled impact demonstration. The multiple ground-controlled take-offs, approaches and landings in the early 1980s demonstrated it is possible to command a four turbine engine commercial aircraft from the ground. Government use of drones in active military theatre only became public during and after the Gulf War's Operation Desert Storm. The live pictures transmitted at the time, of the 'bad guys running away', illustrated clearly to the public the need for weapons platforms, and with subsequent military activities in Afghanistan, Gulf War 2, etc., the deployment of drones armed with missiles was a common theme and occurrence during press briefings.

Boeings' admission, further to the application and grant of a patent to remotely take control (take-over) a commercial aircraft, highlighted that even with armoured flight deck doors and covert Sky Marshals mingling in the passenger cabin with the travelling public, it is still possible for an aircraft to be taken over by *perpetrators* with hostile intensions. Boeing admits that the post-9/11 security improvements had done little to prevent a commercial aircraft being used as a high-value projectile weapon, i.e. flying it into buildings or the ground. Europe responded, around the same time as Boeing made the patent claim, with Airbus announcing that they had been working very closely with BAe Systems and Honeywell on their own landing system, which was *nearing completion*. Any such uninterrupted landing system would require minimal additional equipment, mostly in the LRU within the Electrical and Avionic bay. Fully automated landings (CAT III C) of large commercial aircraft in zero visibility have been made possible by the maturing of this standard technology for many years, based on the instrumental landing systems transmissions that are fitted at most airports. An aircraft that was experiencing difficulty with an active uninterruptable landing system, would not necessarily land at the nearest and most convenient airfield, but rather the designated approved airport identified as being capable of receiving hijacked aircraft. The admission by both major airframe OEMs that an automated ground-controlled landing system was nearing completion was a surprise to the world's media, but was quickly forgotten for many years until the disappearance of MH370 (believed to be pilot-controlled homicide/suicide/sabotage), and two separate events -LAM Mozambique 470 and GermanWings 9525. It has been proven (including evidence from the digital data of the DFDR and DCVR) that suicidal pilots locked their fellow pilots out of the flight deck, and then deliberately crashed the aircraft containing passengers and crews, killing all and destroying the aircraft. Both of these aforementioned 'accidents' have highlighted the need for a ground-controlled landing system, but clearly, such systems were not fitted or worse, were not functioning correctly in these two catastrophic homicidal events. The Malaysian MH370 accident report even cites that the B777 aircraft that disappeared could not have had the Boeing uninterrupted landing system fitted (that would have prevented the occurrence) because the manufacture of the 'disappeared' aircraft was in 2002, four years before 2006 when Boeing received the patent. The report does not mention whether any other aircraft in the fleet have this uninterrupted landing system technology, or whether it is a subscription-based product.

The driving factor for improving commercial aviation, in terms of automated landing systems with covert cameras fitted throughout the aircraft, must not be on the basis of security. This is because there are so few security events or accidents. You cannot benchmark an airline's safety record based on the number of crashes this month/year/decade. Rather, the justification must be financially driven. The need to stream as much live operational data, via satellite communications to ground servers, will allow for real cost savings. The approved flight routes will be monitored much more rigorously, with deviations carefully considered by the airlines and ATCs in real-time. This will improve the fuel burn of each sector, saving money and making significant environmental improvements. The individual component performances, including the bigger system performance evaluations, will allow for better engineering decisions and planning to be conducted, again making significant financial savings. Other simple modifications include the requirement for all commercial aircraft to require a single-use code to start the engines (and fly a pre-agreed sector). This will prevent thefts from ground intruders, as large aircraft are rarely locked, and have no formal code or key-based systems fitted - unlike small 2/4/6 seat aircraft. A chain-link fence, topped with coils of razor wire, is not a sufficient deterrent for intruders as various airport trespass events have demonstrated. The trespass events recorded are rarely seen as active terrorism-based occurrences (with the exception of the Tamil Tiger attack on Bandaranaike Airport, Sri Lanka as discussed in Chapter 7), rather they are miscreant criminal trespass events by local individuals with little or no hostile intentions. Another modification required is removing the ability for pilots to disable the transponder operation in-flight. The ADS-B transponders should be updated with software, thus only when the aircraft is on the ground (weight on wheels); not moving (from GPS data); and the park brake is set to 'ON'; should the transponder system be capable of deactivating (transmission) and be able to switch-off. Other critical systems should be modified in a similar manner to the proposed concept for the transponder: the ECS, the electrical generators, etc. Such systems if switched-off pose a risk to the critical life support systems of the aircraft, and the justification for isolating during 'an unusual' event is ambiguous and outdated, as ground-based control logic should be the only means to switch off such systems. Finally, the circuit breakers for critical systems (including the DFDR, DCVR, communication suite) need to be relocated from the flight deck or electrical and avionic bay to an unpressurised area of the aircraft. This will prevent electrical isolation by pilots in-flight: routine maintenance can still access the C/Bs, but the ability to open circuits in-flight needs to be removed as soon as practicable.

The lack of reported automatic landing events (in the general or industry press) indicates that if such an uninterrupted landing system has been installed, it has not been deployed and used in a hostile context with successful results. This is because passengers would report the landing events immediately on social media which is outside of the direct control of any state: furthermore, the void in technical literature (such as AMM/FCOMs, etc.) implies potentially that the manufacturers and (regulator) states have successfully censored the subject matter. Another more realistic possibility is that while such an automated landing system might exist on paper, the actual deployment across all heavy aircraft fleets has yet to take place, not least as such a modification is not a mandated modification by ICAO or the regulators.

While the justification to fit an uninterruptable landing system is clear, if this new system were fitted and became operational, once the occupant realises that the aircraft will land and the security authorities will meet/storm the aircraft, there is a realistic potential for further acts of subterfuge and sabotage. In-flight fires in commercial aircraft have caused many passenger flights to crash over the years. Fire remains one of the biggest risks in the enclosed pressurised environment of the cabin. Toilets remain a real problem (including the paper inside), and the single ceiling detection system fitted is often tampered-with by desperate passenger smokers who go to great lengths to smoke in these enclosed spaces. If passengers are willing to interfere with such systems just to smoke a cigarette (risking a catastrophic fire), a more determined individual in the private confines of the toilet would have little difficulty starting and establishing a cabin fire, resulting in flash over some 4 minutes later. Likewise, individuals in the 'secure' locked flight deck have access to the electrical and avionic bays, usually located underneath the pilots. In such an automated landing scenario, it would be important for the armoured trap door to be locked shut, preventing unauthorised access by any occupant. Likewise, the flight deck and forward cabin floor pose an additional security risk, being manufactured from lightweight, composite honeycomb materials. These floorboards need to be reinforced with Kevlar within the panels (to prevent ballistic penetration), and the floors secured in such a way that no person can penetrate the cabin floor into the cargo bay or electrical and avionic bay.

Finally, the data security of such an uninterrupted landing system is a paramount feature that will underpin any potential deployment. As both Boeing and Airbus already have considerable expertise in AES military-grade encrypted communications, used by militaries, intelligence services and government signals agencies, it would seem highly unlikely that a rogue state or hacker would be able to take control of a flying aircraft and force it to land. Providing the ultimate use of such a system remains within the expertise of the aircraft OEM, the risk of a hostile ground-based digital take over appears much less likely.

Index

A

AAI Corporation & Israel Aircraft
    Industries RQ-2 Pioneer drone,
    160, 161
acceptable deferred defect, 54
active failure (Swiss cheese), 85
advanced encryption standards, 175
Air Accident Investigation Branch (UK),
    72, 95, 176
Airbus predictive maintenance, 67–69
aircraft communication addressing and
    reporting system, 5, 61–63, 67, 70
aircraft maintenance manual, 54, 165, 180
air data computer, 36
air traffic control, 1, 36, 80, 117, 135, 139,
    144, 171, 174
Ariana Afghan Airlines flight 805 hijack
    and siege, 165, 166
automatic dependent surveillance
    broadcast, 119, 171, 180
autonomous taxi, take-off and landing
    system, 9, 166
avionic bay need for additional
    protection, 176–178, 180, 181

B

Bandaranaike Airport attack, 6, 98–100
base deferred defect, 54
bathtub curve for reliability, 55
Boeing B737-maxx MCAS crash related
    events, 168, 169
Boeing B-17 remote control
    aircraft, 155, 156
Boeing B-720 remote control aircraft for
    fire and crash test (FAA), 159,
    160
British Airline Pilot Association, 72, 73
British Airways flight 2069 B747-400
    security event inflight, 95–98

British European Airways flight BE548
    Hawker Siddeley Trident crash,
    72–75, 77
British Midland flight 92 B737-400 crash,
    94, 95
British Overseas Airways Corporation,
    18, 23, 24
Bureau d'Enquêtes et d'Analyses, 137,
    139

C

Cambridge Analytica, 69
Cathay Pacific Airways DC3 Betsey, 14,
    15
Cessna 172 simplified flight controls, 34,
    35
China Southern Airlines, B737-300
    accident, 30
Circuit breakers and locations, 170
Civil Aviation Authority, 1, 27, 89
cockpit voice recorder (analogue), 23, 27,
    29, 30, 107, 111
Consolidated PBY Catalina Miss Macao
    hijack, 14
continual descent approaches, 174
COVID-19, 148–150
crew/cockpit coordinated concept, 83,
    84
crew resource management, 84, 85, 95
Cubana de Aviación Douglas DC3
    hijack, 17

D

data security and encryption for
    uninterrupted landing systems,
    175, 176
de Havilland Comet DH106 multiple
    accidents, 23

de Havilland Queen Bee remote control
    aircraft, 155, 156
digital cockpit voice recorder, 111, 113,
    135, 139, 179, 180
digital flight data recorder, 7, 30, 31, 32,
    57, 70, 112, 115, 117, 133, 135,
    144, 179, 180

E

Egypt Air flight 990, 114–116
emissions trading systems, 174
end 2 end encryption, 175
English Electric Canberra high altitude
    remote controlled aircraft, 157
enterprise-wide system, 64
environmental control system, 4, 7, 9, 32,
    38–40, 44, 46–49, 122, 132, 149,
    150, 169, 180
Ethiopian Airlines flight 961 hijack, 19
European Aeronautic Defence And
    Space, 166
European Union Aviation Safety
    Agency, 52, 90, 150
exhaust gas temperature, 57

F

Federal Aviation Administration, 8, 16,
    17, 56, 101, 158
federal aviation requirements, 53
Federal Bureau of Investigation, 8, 143,
    144
fire, 19, 27, 31, 46, 47, 81, 84, 85,
    95, 98, 146–148, 154, 158, 176,
    180, 181
First World War remote control aircraft,
    154, 155
flash over occurrence (fire), 98
flight crew operating manual, 38, 165,
    180
flight controls, 4, 32, 33–36, 77, 113, 165
flight data recorder (analogue), 27,
    29–31, 107
flight deck door, 6, 7, 84, 93, 95, 98, 101,
    103, 109, 111, 132, 136, 138, 140,
    145, 164, 177

flight management computer, 4, 119,
    171, 176
Ford tri-motor hijack, 12
full authority digital engine control, 5, 60

G

Germanwings flight 9525, 136–142
global positioning system, 108, 109, 164,
    170, 171, 180
global system for mobile, 60
ground proximity warning systems, 164
General Atomics Predator drone, 161, 162
Gulf war operation desert storm, 153

H

homicide (inc. attempt), 3, 7, 13, 21, 96,
    101, 103, 108, 109, 111, 113, 116,
    133, 135–137, 142, 146, 148, 151,
    154, 165, 168–170, 173, 175, 179
Horizon Air DHC dash 8-Q400 theft
    event and deliberate crash,
    142–145
hypoxia, 4, 38, 47, 122, 132, 176

I

inertia navigation systems, 36, 106, 108
International Civil Aviation
    Organisation, 2, 30, 90, 102, 119,
    135, 180
instrument landing system, 9, 36, 37,
    154, 165, 167
Iran Air Airbus A300-200 flight 655
    shoot down by USS Vincennes,
    108

J

Japan Airlines flight 351 hijack, 18

K

Korean Airlines flight KL 007 deliberate
    shootdown by Soviet military,
    105–108

L

LAM Mozambique flight 470, 135, 136
Latent failure (Swiss cheese), 85
line replaceable unit, 46, 176, 178
liquids (security restricted carry on
        items), 6, 93, 103
Lockheed C130 platform to launch
        remote controlled
        aircraft, 157, 158
Lockheed Skunk Works U-2 Dragon
        Lady ECS and flight suit, 48
London 7–7 terrorist attacks, 102, 103
London Gatwick airside intrusion
        security event, 144, 145
Luftfahrt Bundesamt, 8, 138

M

Malaysian Airlines flight MH370
        missing presumed hijacked,
        117–133
Malaysian Airlines flight MH17 shoot
        down by presumed Russian
        military, 133, 134
master minimum equipment
        list, 53, 54
maximum take-off mass, 1, 38, 145
mean time to failure, 55, 56
metal detectors, 18, 21
minimum equipment list, 53, 54

N

national aviation authority, 1, 52, 53, 54,
        91, 142, 172
National Transportation Safety Board,
        111–113, 115, 116, 143, 147
9–11 terror attacks using aircraft, 6, 21,
        59, 100–102, 104, 137
Northrop Grumman RG-4 Global Hawk
        drone, 162, 163

O

one time use codes proposed for engine
        start authority, 171

original equipment manufacturer, 52–54,
        56, 60, 64, 65, 67, 69, 113,
        171–173, 175, 176, 178,
        179, 181
outflow valves, 47

P

Pacific South West Airlines flight 1771,
        109, 110
Palantir predictive maintenance, 67–69
plate fin heat exchanger, 38, 41, 42, 44
Piper cub homicide event, 13
Popular Front for the Liberation of
        Palestine at Dawson's Field, 18
'power by the hour' engine pricing, 59
pressure altitude, 46

Q

'Q' feel for flight controls, 35
quick access recorders, 57, 70

R

radio altimeter, 165
ram air turbine, 121
remotely piloted vehicles, 8, 159
reliability, 51–69
Rolls Royce commercial engines
        predictive maintenance, 64–67

S

sabotage, 6, 85, 90, 154, 176, 179, 180
safety management system, 6, 71, 89, 90,
        91, 173
Saudia Airlines flight 163 Lockheed
        L-1011 inflight fire, 84, 85
search and rescue, 28, 107, 122, 136
security screening, 2, 3, 19, 85, 103,
        111, 171
Senator George Smathers
        (Florida State), 17

severe acute respiratory syndrome
    coronavirus 2 (covid 19), 149,
    150
Silk Air flight 185, 111–113
Sikorski S38 hijack, 12
sky marshals, 6, 16, 21, 93, 101, 104, 136,
    164, 178
Skywise predictive maintenance, 67–69
software, hardware, environment,
    liveware, liveware model, 5, 77,
    78, 81, 84
Special Air Services, 166
suicide (inc. attempt), 6–8, 93, 96, 97,
    100–102, 105, 109, 111, 113, 116,
    136–138, 140, 142, 144, 147, 150,
    179
Swiss cheese model, 6, 85–88, 91

T

Tenerife accident, 78–80
terrorist/terrorism, 3, 12, 21, 93, 99, 100,
    101, 103, 104, 136, 143

Toilet smoke detectors, 98, 146–148
transponders and proposed
    modification, 170
trespass, 145, 146, 179

U

underwater locator beacons, 28
uninterrupted landing system, 154, 164,
    175, 176, 178–181
Unmanned Aerial Vehicle, 161–163,
    167
unscheduled maintenance, 53

W

Warren's "Red Egg" flight recorder, 26
Wright brothers Kittyhawk, 34

X

x-ray screening, 2, 3, 18, 19, 21, 101

Printed in the United States
by Baker & Taylor Publisher Services